AF528597

The book on
"Butterflies in Assam: Diversity and Ecology"
is dedicated to my parents

Late Nilima Bora
&
Late Jnanendra Lal Bora

Dipsikha Bora

er Grass Blue, *Zizina otis* Fabricius

Contents

A newly emerged Sergeant

CHAPTER I

The butterflies

Evolutionary background

Dipsikha Bora

The conspicuousness and attractiveness of the magnificent lepidopterans have made the lepidopteran study, a subject of interest for amateur entomologists, naturalists, insect systematists, and ecologists. They are well placed in the works of cartoonists, poets, nature writers, story writers, artists, painters, and craft designers. Academic interests concerning butterfly studies are mostly directed toward aspects of diversity, evolution, climate change effects, pollutant effects, and ecological roles. Butterflies constitute a diverse group of insects belonging to the order Lepidoptera. The latter is one of the most studied insect group with around 1,60,000 described species [1][2], 135 families, and 45 superfamilies [3]. The globally described lepidopterans are said to constitute about 10 % of the described living organisms and a major contributor of 2.5 billion(approx.) natural history collections across the globe [3]. Out of the four Lepidopteran suborders, Zeugloptera, Dacnonypha, Monotrysia, and Ditrysia, the nonditrysian groups are paraphyletic while ditrysian groups are reported as monophyletic (ambiguity with Tineoidea). Ditrysia is the most speciose group and includes the charismatic butterflies belonging to two superfamilies, Papillonoidea and Hesperioidea. The Papillonoidea includes five families, Nymphalidae, Papillonidae, Pieridae, Lycaenidae, Riodinidae and Hesperioidea includes only one family, Hesperidae. Papillonoidea and Hesperioidea are also included by some authors under Rhopalocera. Nearly 19,000 species of butterflies have been described so far [4]. Recent global phylogenetic approaches using genetic sequences indicate the need for the reclassification of some butterfly tribes [5].

The evolutionary perspective of butterflies deals with the origin of the history of butterflies and how those historical events took shape. The perspectives in the chapter are discussed in terms of fossil history, coevolution, and convergent evolution.

Fossil history: The study of fossil insects is known as Paleoentomology. Such studies revealed that the primitive insects were larger in size as compared to the modern insects. Insect remains are mostly preserved due to the presence of chitin between hard surfaces, amber, sedimentary rocks, stones, etc. The earliest hexapods having a common ancestor with crustaceans probably arose in the Silurian. During the Silurian period the first land plants appeared in the fossil record, and insects probably colonized land shortly afterwards. Well-preserved insect fossils are available from the Early Devonian and hence insects are said to have originated 412 million years ago, during the Early Devonian Period. The earliest known insect fossil is *Rhyniognatha hirsti*. The winged insects first appeared in the Carboniferous period (350 million years). They are postulated to have risen much earlier (Early Devonian) because the earliest fossils of winged insects had well-developed and complex wings (Misof et al.,2014). The dragonflies, one of the living fossil groups cannot fold their wings. Wing folding evolved in the Carboniferous period, after the evolution of wings. These wing-bearing pterygotes underwent a major divergent radiation in the carboniferous period (356 to 299 million years ago) which is known for its wet, warm climates and extensive swamps. The phenomenon of complete metamorphosis evolved sometime late in the Carboniferous or early Permian. The endopterygotes underwent another major radiation in the Permian (299 to 252 million years ago). At the end of the Permian, the biggest mass extinction in history occurred, collectively called the Permian–Triassic extinction event which might have paved the way for major radiation, especially in the Holometabola in the Cretaceous [6].

The fossil history along the geological timescale traces the origin of lepidopterans back to *Archaeolepis mane* Whalley (Archaeolepidae), a record from the lower Jurassic period (193Mya) [7]. This fossil is a kind of calcareous flatstone deposit and reflects wing venation pattern similar to trichopteran, a sister order of Lepidoptera [8]. Fossil records have been revealed from compression-impression, amber-copal, asphaltam-tar sands, peat-lignite, salt, gut content-coprolite of insectivorous animals, and trace deposits [7]. From the lower Jurassic to the Holocene epoch of the Cenozoic Era, a total of 4561 lepidopteran fossils have been detected. A total of 985 lepidopteran fossils have been assigned to 23 extant lepidopteran superfamilies out of which 142 have been assigned to Papilionoidea. The first reported fossil of Ditrysia is a leaf mine from the mid-Cretaceous (103 Mya) which is assigned to the modern family, Gracillaridae [9].

Coevolution: The inferred age of Papilionoidea and its constituent family is said to be 140-80Mya (lower to middle cretaceous) [4]. Study of insect-mediated feeding traces suggests the occurrence of major lepidopteran lineages to gymnosperm dominated late Jurassic period [10]. The diversification of Angiosperms and the overwhelming abundance over gymnosperms in the Cretaceous period marks one of the greatest terrestrial radiation which was referred as the "Abominable mystery" by Darwin. The mystery was expressed in the letter Darwin wrote to his friend J.D. Hooker in 1879 as " .....the rapid development as far as we can judge of all the higher plants within recent geological times" [10][11]. Seed dispersal and pollination by diverse animals also contributed to angiosperm dominance.

Coevolution is an arms race and involves a reciprocal genetic change in interacting species resulting from natural selection. The simplest form is **specific coevolution** expressed phenotypically. Mouth part diversification on insects tailored to feed on different food resources serves as a bridge for insect-plant association either through pollination or herbivory and is presumed to have triggered the radiation of flowering plants (120-70Mya). Charles Darwin once while inspecting an orchid having a long nectar spur wondered "which insect can suck in" the nectar and years later a giant hawk moth (Sphingidae) was discovered with a long proboscis (9.5 inches) in Madagascar that could easily touch the nectar pooled at the base of the nectary. The orchid was *Angraecum sesquepedale* and the moth was *Xanthopan morgani praedicta*. This is a classic example of specific coevolution. Studies on the extant butterflies reveal variations in the length of butterfly proboscis which can be related to the corolla length of the flowers they visit.

The co-radiation of angiosperm and plant-dependent insects during the cretaceous period starting from 130Mya is postulated and lepidopterans probably diversified in concert with the angiosperms. The evolutionary divergence of yucca moths can be referred to here as an example of specific coevolution in which, host plant-based evolutionary change in the structure of genitalia led to diversification [12][13]. While species interactions shape evolutionary traits, the influence of within-species interaction is not to be ignored where processes like sexual selection and sexual conflict are widespread. Coevolutionary changes in genitalia due to specific interaction of male and female lepidopterans during reproduction is another example and is under extensive study [14]. A study on long-term interaction of the Saturniid moth, *Antheraea assamensis* with its host plants, *Persea bombycina* and *Litsea monopetala* revealed induction of chemical defense response in the host plants through herbivory. The plant chemicals in turn exerted inhibitory action on the digestive protease of the lepidopteran larvae [15]. In **escape and radiate coevolution**, a reciprocal adaptation is involved.

The day-flying butterflies are said to have evolved from the nocturnal moth ancestors about 80 MYa in the late Cretaceous period [2]. Controversies prevail. Did they evolve to enjoy the exuberance of nectar of the angiospermic flowers during broad daylight or from the selection pressure posed by predatory bats [16] is a question pondered at.

The association of butterflies and their host plants also formed the foundation of the coevolutionary theory of Ehrlich and Raven (1964). In coevolutionary studies, patterns of interactions between two groups of organisms having close ecological relations are examined. Insect-host plant interactions provide the platform for such studies where consequence and influencing factors develop one after the other. Ehrlich and Raven mined out a considerable amount of information on butterflies and their larval host plants with regard to the factors influencing the preferences for a particular food plant, the structural and mechanical characters of the plants, the chemical factors, nutritional values and repelling or attracting factors. For instance, Ehrlich and Raven analyzed the preference of host plants by the different subfamilies of the Papilionid butterflies. The subfamily Baroniinae feeds on *Acacia* (Leguminosae). Several genera of the subfamily Parnassiinae feed on plants under Aristolochiaceae. Under the subfamily Papilioninae comprising three tribes, several genera of tribe Troidini feed on Aristolochiaceae, genera under tribe Graphinii prefer families like Lauraceae, Annonaceae, Magnoliaceae and other allied families referred to as Woody ranales. The members of Papilionini feed on Rutaceae, Umbelliferae, Leguminosae, and many others. Based on the host plant preference they concluded that Baronia (Baroniinae) does not feed on Aristolochiaceae and represents a phylogenetic line that diverged early from the rest of Papilionidae. Later studies by Zanz and Nylin (1998) using new phylogenetic evidences corroborated the finding of Ehrlich and Raven (1964), that related butterflies feed on related plants. They also reaffirmed that the pattern of association during insect-host plant interaction is dependent on host shifting among closely related hosts. Plants can undergo diversification and insects may manage to adapt to a diversified plant resulting in host shifting and colonisation. Rather than entering into the process of speciation, they may even prefer ancient hosts for colonization. Such a condition where ancient character remains unchanged is explained by **phylogenetic niche conservation**. If a gene exchange occurs among individuals occupying an ancestral niche indicating a specific microhabitat, and those inhabiting a new niche and there is a fitness competition of character states, selection will prefer the ancestral character states as most of the populations occupied the ancestral niche [17].

Convergent evolution: Convergent traits are often adaptations of different lineages evolving independently to similar environmental conditions. Natural selections may not always lead to evolution. But adaptations evolving by natural selection causes evolution which is the central

Butterflies IN ASSAM

DIVERSITY AND ECOLOGY

Written and edited
by
Dr. (Mrs) Dipsikha Bora

Contributing authors
Anisha Phukan, Phalgun Chetia, Amrita Mech

2025
Regency Publications
A Division of
Astral International Pvt. Ltd.
New Delhi – 110 002

© 2025, AUTHORS

ISBN: 978-93-5919-753-1 (Hardback)

Publisher's Note:

Every possible effort has been made to ensure that the information contained in this book is accurate at the time of going to press, and the publisher and author cannot accept responsibility for any errors or omissions, however caused. No responsibility for loss or damage occasioned to any person acting, or refraining from action, as a result of the material in this publication can be accepted by the editor, the publisher or the author. The Publisher is not associated with any product or vendor mentioned in the book. The contents of this work are intended to further general scientific research, understanding and discussion only. Readers should consult with a specialist where appropriate.

Every effort has been made to trace the owners of copyright material used in this book, if any. The author and the publisher will be grateful for any omission brought to their notice for acknowledgment in the future editions of the book.

All Rights reserved under International Copyright Conventions. No part of this publication may be reproduced, stored in a retrieval system, or transmitted in any form or by any means, electronic, mechanical, photocopying, recording or otherwise without the prior written consent of the publisher and the copyright owner.

Published by: **Regency Publications**
A Division of
Astral International Pvt. Ltd.
– ISO 9001:2015 Certified Company –
4736/23, Ansari Road, Darya Ganj
New Delhi-110 002
Ph. 011-43549197, 8130496929
E-mail: info@astralint.com
Website: www.astralint.com

Photographs: (Technical) Dr. (Mrs) Dipsikha Bora
Phalgun Chetia, Amrita Mech, Pooja Nag, Niranjan Baruah, Dulen Bora, Anisha Phukan, Jhuma Gogoi, Jyoti Agrahari, Anjan Chaliha and Longnit Terang

Cover Photo: Plains Cupid, *Chilades pandava* Horsfield
Photograph: Jhuma Gogoi

Design & Layout: Himangshu Lahkar

Printed at: Replika Press Pvt Ltd.

Preface

"The Butterflies in Assam: Diversity and Ecology" is structured into five chapters, each encompassing different aspects of the lives of butterflies. Chapter I will provide insight into the evolutionary history of butterflies within a broader context of insect evolution. Chapter II will give information on structural details. Understanding the structural features of butterflies is essential for accurate identification and appreciation of their diversity. Chapter III will contain more than 165 photo plates of butterflies found in Assam with field characters. By highlighting the unique characteristics of each species, this chapter serves as a practical guide for butterfly enthusiasts and researchers alike. Chapter IV will describe the ecological importance of butterflies and Chapter V will introduce the methods of studying butterflies.

The state of Assam is located in the northeastern corner of India. Northeast India comes under Eastern Himalayan bio-diversity hot spot region and harbours rich biodiversity. About fifty percent of butterflies available in India are said to be present in the North east. The habitats within the political boundary of Assam include river valleys, dense tropical rainforests, remnants of forests, forest treks, grass lands, gardens, and parks. The butterflies need different host plants for different stages of their life and hence the rich diversity of plants supports the abundance of butterflies across the state. The adults need rain water pools also for obtaining nutrients from the soil which is naturally available. The average rainfall in the state is about 2000 mm and usually highly deficient period is rare. Butterfly collection and their taxonomic identification in India have a long history starting with the colonial Britishers and the Bombay Natural History Society has been associated from the beginning. EH. Aitken, WH. Evans, MA. Wynter-Blyth, Isaac Khemikar, K. Kunte and M. Haribal are the pioneers in the field. In their writings and also in the present butterfly diversity portal of NCBS (National Centre for Biological Sciences), butterflies from north east India are incorporated. Haribal in 1992 reported butterflies from administrative areas of Sikkim located in North East India. Educational institutions of higher studies in Assam reported from time to time, presence of good number of butterflies in different habitats and altitudes across Assam and the adjoining states.

The book offers a comprehensive guide to the diverse butterfly species inhabiting Assam's gardens, parks, and forests, making them accessible to all. The butterflies preferring forest habitats were photographed from fringe areas of Dehing Patkai National Park, Jakai Reserve Forest, Kaziranga National Park, Ultapani Reserve Forest and Gibbon Wildlife Sanctuary. The author's personal experiences of exploring Assam's various regions throughout different seasons enrich the book's content, providing valuable insights into butterfly behaviour and distribution. The field characters provided for identification cater to a wide audience including undergraduate and post graduate students of Zoology, researchers, and citizen scientists. The hierarchial ranking of life forms and phylogenetic relationships may take new shape with the advances in molecular tools. Extensive taxonomic restructuring is in progress and appears dynamic. Therefore, morpho-taxonomic characters are regaining importance for revisiting classification of fauna and flora using molecular tools. The butterfly is an important indicator of environmental changes including effect of climate change. Given the paucity of publications specifically addressing the field identification of butterflies in urban and forested areas of Assam, "The Butterflies in Assam: Diversity and Ecology" fills an important gap in the literature. Its practical approach and user-friendly format are expected to be well-received by nature lovers, students, entomologists, amateurs and environmentalists alike. The readers may correspond to the author at E mail id. dipsikhabora03@yahoo.com for suggestions facilitating improvement in future editions.

Dipsikha Bora

Acknowledgement

To enrich my learning experience in entomology and insect ecology, I started my journey on observation of a magnificent creature on earth, the butterflies, since 2009. Over the years, I gathered data intermittently for recording their diversity in the urban areas of Assam. Earlier I published a research paper on "Matrix habitat use by Butterfly communities in the vicinity of Jakai reserve forest of Assam" in a Journal published from the Department of Life Sciences in 2011. Many individuals contributed towards collection of data and the final compilation into this book to whom I shall always remain grateful.

I am thankful to Anisha Phukan, a Ph.D. student under Dibrugarh University who being one of the contributors, took keen interest in helping me compiling the information in the form of a book. Phalgun Chetia and Amrita Mech contributed by filling the gaps in my collections of butterfly diversity and I appreciate their moral and material support in bringing out the book. My thanks also go to Pooja Nag, a Master's student, batch 2022 whose contributions extended my range of collection to the fringe areas of Dehing Patkai National Park, Assam. I am thankful to those students of Entomology course who from time to time enthused me to include some of their collections as a repository in the Department of Life Sciences, Dibrugarh University.

Over the years while pursuing the work, a young friend with a keen interest in photography, Dulen, Chauffeurs Jintu, Sahidul and Pranjit, accompanied me in my survey and I owe my thanks to them for keeping patience for long waiting periods.

I am indebted to my husband, Niranjan Baruah for his logistic and moral support during the field surveys which I carried out during holidays and long vacations. Special thanks to Longnit Terang, Anjan Chaliha, Lengdon Handique, Jhuma Gogoi for providing a few photographs. I am thankful to my children for their love, support and encouragement in my academic ventures. Thanks to all my friends and well wishers whose encouragement was vital in bringing out the book.

I wish to offer my thanks to Himangshu Lahkar for designing the book and rendering it into a publishable form.

Finally, I owe my gratitude to the Publisher, Astral, New Delhi for kindly taking the responsibility of publishing this work .

Dipsikha Bora

Papilio polytes

dogma of the theory of evolution. The peppered moth, *Biston betularia* in England was characterized by black marks on the wings giving the appearance of peppered surface. In mid-1800, when England was passing through industrial revolution, melanic mutation in peppered moth characterized by darkened wings suddenly spreaded. During this period of industrial revolution vast quantities of coal was burnt producing large amount of soot leading to darkening of tree trunks. Moths with peppered appearance became conspicuous rendering them highly vulnerable to the predators. Melanic allele then came into existance leading to successful camouflaging of the melanic moth. But their population decreased between 1960 to 2000 when coal burning reduced. The rise and fall of melanic mutation cites a good example of natural selection and evolution in real time [17]. Mimicry observed among butterflies is a kind of convergent evolution. Mutations having a large effect can cause such evolution. Within the genus *Heliconius*, structural colour patterns have evolved multiple times involving iridescent, non-iridescent, hybrid colors, or different degrees of iridescence. Genetic analysis of the structure colour pattern of *Heliconious* butterfly co-mimetics, *Heliconius erato* and *Heliconius melpomene* showed multiple genes to be involved in the trait of colour and structure and a different trajectory for the evolution of structural colour [18]. Convergent evolution between mid-mesozoic poorly preserved kalligrammatid lacewing (Neuroptera) and modern butterflies (Nymphalid) have also been detailed in terms of conversion of simple eye spots to complex eye spots with concentric circles of pigmented cells used for deflection of predators [19].

Finally, I conclude by stating that, evolutionary history is never complete. There may be many more factors behind the evolution of the novelty and diversity of butterflies. It always leaves a scope for another review.

References

1. Kristensen, N.P., Scoble, M.J., and Karsholt, O. (2007). Lepidoptera phylogeny and systematics: the state of inventorying moth and butterfly diversity Zootaxa 1668: 699-747 (2007) www.mapress.com/zootaxa/ In: Zhang, Z.-Q. & Shear, W.A. (Eds) (2007) Linnaeus Tercentenary: Progress in Invertebrate Taxonomy. Zootaxa, 1668, 1-766
2. Kawahara, A. Y. et al. (2019). Phylogenomics reveals the evolutionary timing and pattern of butterflies and moths. Proc. Natl Acad. Sci. USA 116, 22657-22663

3. Prins, J.D. (2017). Lepidoptera Collection Curation and Data Management. Intech Publisher http://dx.doi.org/10.5772/intechopen.70925
4. Mitter, C., Donald, R.D. and Michael, P.C. (2017). Phylogeny and Evolution of Lepidoptera. Annu. Rev. Entomol. 62,265–83
5. Kawahara, A.Y., Storer, C., Carvalho, A.P.S. et al. (2023). A global phylogeny of butterflies reveals their evolutionary history, ancestral hosts and biogeographic origins. Nat Ecol Evol **7**, 903–913 https://doi.org/10.1038/s41559-023-02041-9
6. Misof, B., Liu, S., Meusemann, et al. (2014). Phylogenomics resolves the timing and pattern of insect evolution.Science. 346 (6210),763-767. DOI: 10.1126/science.1257570
7. Sohn, J.C., Labandeira, C., Davis, D.R. (2015). The fossil record and taxonomy of butterflies and moths (Insecta, Lepidoptera): implications for evolutionary diversity and divergence-time estimates. BMC Evol. Biol. 15:12, 1-15
8. Grimaldi, D.A., and Engel, MS. (2005). Evolution of the Insects. Cambridge, UK: Cambridge Univ. Press
9. Labandeira, C.C., Dilcher, D.L., Davis, D.R., Wagner DL. (1994) Ninety-seven million years of angiosperm insect association: paleobiological insights into the meaning of coevolution. PNAS 91,12278–82
10. Berendse, F. and Scheffer, M. (2009). The angiosperm radiation revisited, an ecological explanation for Darwin's abominable mystery. Ecology Letters. 12, 865–872
11. Soltis, P.S., Folk, R.A., Soltis, D.E. (2019). Darwin review: angiosperm phylogeny and evolutionary radiations. Proc. R. Soc. B 286: 20190099. http://dx.doi.org/10.1098/rspb.2019.0099
12. Althoff, D.M. and Segraves, K.A. (2021). Evolution of antagonistic and mutualistic traits in the yucca-yucca moth obligate pollination mutualism. Journal of Evolutionary Biology. https://doi.org/10.1111/jeb.13967
13. Althoff, M.D. (2013). Shift in egg-laying strategy to avoid plant defense leads to reproductive isolation in mutualistic and cheating yucca moths. Evolution. 68-1, 301–307
14. Cordero, C. and Eberhard, W.G. (2005). Interaction between sexually antagonistic selection and mate choice in the evolution of female responses to male traits. Evolutionary Ecology. 19, 111–122 DOI: 10.1007/s10682-004-7918-2
15. Mech, A., Choudhury, K.K. and Bora, D.S. (2021). Lipoxygenase - trypsin inhibitor activity axis induction in the host plants of muga

silkworm, *Antheraea assamensis* Helfer by feeding. Entomological Research. doi: 10.1111/1748-5967.12543

16. Yack, J., Fullard, J. (2000). Ultrasonic hearing in nocturnal butterflies. Nature **403**, 265–266 https://doi.org/10.1038/35002247
17. Futuyma, D. and Kirkpatrich, M. (2018). Evolution. Fourth edition, Oxford University Press
18. Brien, M.N. et al. (2022). The genetic basis of structural colour variation in mimetic Heliconius butterflies. Phil. Trans. R. Soc. B 377: 20200505. https://doi.org/10.1098/rstb.2020.0505
19. Labandeira, C.C., Yang, Q., Santiago-Blay, J.A., Hotton, C.L., Monteiro, A., Wang, Y.J., Goreva, Y., Shih, C., Siljeström, S., Rose, T.R., Dilcher, D.L., Ren, D. (2016). The evolutionary convergence of mid-Mesozoic lacewings and Cenozoic butterflies. Proc Biol Sci. 10;283(1824):20152893. doi: 10.1098/rspb.2015.2893. PMID: 26842570; PMCID: PMC4760178.
20. Ehrlich, P.R. and Raven, P.H. (1964). Butterflies and plants: A study in coevolution. Evolution 18 (4), 586-608

Phanerozoic Eon							
Paleozoic Era			Mesozoic Era			Cenozoic Era	
Silurian (438-410)	Devonian (410-360)	Permian (286-248)	Triassic (248-208)	Jurassic (208-145)	Cretaceous (145-65)	Tertiary	Quaternary (1.6- present)
							Extant Lepidopterans Extant Gymnosperms Extant Angiosperms
						Angiosperm diversification Butterfly fossils (30-40 Mya)	
					After mass extinction, gymnosperm declined, and angiosperms diversified. Radiation of angiosperm pollinating insects (125-92 Mya) Papilionid butterflies (80 Mya)		
				Gymnosperm dominated late Jurassic			
			Angiosperm fossil pollen recorded (240-247 Mya) Major radiation of holometabolous insects				
		Gymnosperm structures established Evolution of metamorphosis Winged insects evolved					
	Gymnosperms evolved (seed). Appearance of fossil insect, Rhyniognatha hirsti (400 Mya)						
Hexapod evolved from a common ancestor with crustacean							

Lepidopteran evolution in geological timescale

CHAPTER II

Butterfly Morphology

Anisha Phukan and
Dipsikha Bora

Butterfly species are identified by a number of unique features, including a pair of clubbed or clavate antenna, patterns of arrangement of colourful scales covering the body, wings, structure of legs, wings, genitalia, host plants and flight pattern. Mating and egg-laying typically take place near or on the host plants for the larvae. They have acquired the evolutionary landmark in the evolutionary history of insects which is complete metamorphosis. In the case, immature forms are entirely different from their adult forms. A sexually developed adult emerges once the transition is complete. The stages in the life cycle are egg, larva, pupa and adult.

1. Eggs

Butterfly eggs, scientifically known as ova, are the initial stage of a butterfly's life cycle. The eggs can be categorized into two forms, flattened or ovoid, with a horizontal long axis, and upright eggs with equal or longer vertical axes. They vary in size, shape, and color, depending on the species ranging from a few hundred micrometers to a few millimeters in diameter. The shape can be spherical, oval, fusiform or elongated. The eggs are usually attached to a substrate, such as leaves, stems, or other surfaces, laid singly or in batches. The egg contains microscopic pore called as micropyle with canals radiating into the inside of the egg through which sperm can enter during fertilization. The micropyle is located either in a depression at one end of the horizontal axis of ovoid eggs or at the summit of upright eggs. Butterfly eggs have a protective outer layer called the chorion which is a thin and translucent shell that encloses the developing embryo composed of multiple layers, providing mechanical strength and protection against desiccation and physical damage. The chorion often exhibits a textured or sculptured surface, which may be smooth, ridged, pitted or chambered. These surface characteristics vary across species and can aid in identification [1].

2. Larva

The larvae, also known as caterpillars, represent the juvenile stage in the life cycle of butterflies. With a long and segmented body, the larva possesses a well-defined structure consisting of head, thorax, and abdomen. The head features simple eyes, antennae, and powerful mandibles used for feeding. The thorax bears three pairs of true legs and equipped with small claws or hooks, facilitating gripping the surfaces and subsequent movement. Additionally, caterpillars may possess several pairs of fleshy prolegs along the abdomen, each adorned with tiny hooks called crochets that aid in grasping . The insect larvae which resemble caterpillars are called as eruciform larvae. The larvae are enveloped in a soft, flexible skin known as cuticle. This thin outer layer not only protects the caterpillar's internal organs but also serves as a site for muscle attachment and showcases various patterns, colors, and textures that also contribute to camouflage, warning signals, or mimicry. The dorsal surface of the larvae may be smooth or hairy or contain series of tubercles or verrucae with bristles. They may or may not contain anterior or posterior horn. Some species even possess spinnerets, which produce silk used in forming cocoon mostly using leaves or securing pupae by fastening. Caterpillars play a crucial role in the life cycle of butterflies, acting as voracious eaters. This stage allows them to accumulate the necessary nutrients and energy vital for their subsequent transformation into pupae and, ultimately, into the magnificent adult butterflies.

The internal anatomy of butterfly larvae, showcases adaptations crucial for their growth and development. The body cavity houses its digestive system, reproductive organs, nervous system, circulatory system and specialized structures for respiration. The digestive system consists

Larvae of Castor Butterfly

of mouthparts, an alimentary canal with the buccal cavity, esophagus, stomach, and intestine. Enzymes within the gut aid in breaking down plant material to extract nutrients. The respiratory system includes spiracles located on the lateral sides of the body segments, providing openings for oxygen uptake, which then leads to a network of tracheae delivering oxygen directly to the tissues. The nervous system comprises a brain located in the head region and a ventral nerve cord running the length of the body, with interconnected ganglia and lateral nerves coordinating various physiological processes. Malpighian tubules and hindgut-rectum complex, functioning as excretory structures eliminate waste products and maintain fluid balance. Additionally, fat bodies which act like vertebrate liver serve as energy reserve and are involved in synthesizing essential substances required for growth and development. The sternum of prothoracic segment may contain 'osmeteria', a protrusible sac that thrust out to exhale distinctive odour or an internal sac capable of discharging a jet of spray for defensive purpose.

3. Pupa

The pupae of butterflies, scientifically known as chrysalis, is a pivotal stage in their life cycle. Pupae are the non-feeding stage. The food stored during the larval feeding stage is utilized for differentiation and transformation to the adult stage. The pupae may be variable in form. They may have lateral or medial cephalic projections and externally visible or invisible wing pads . They may remain inside a puparium too. Pupae are obtect and adecticous. Often attached to a substrate, such as a branch or leaf, the pupa secures itself through structures like the cremaster, middle silken on varies among butterfly species, ranging from subtle shades of brown, green, or gray to more vibrant hues. Some pupae exhibit patterns or markings that aid in camouflage or mimicry, blending seamlessly with their surroundings for enhanced protection. The pupal stage is primarily a resting phase, where the butterfly appears dormant from the outside. The pupa takes the form of a protective casing or shell, characterized by an oblong or oval shape. It possesses a firm yet flexible outer covering, known as the pupal exoskeleton, which is formed from the hardened cuticle of the last larval instar. Within this casing, the caterpillar undergoes an extensive internal reorganization and new developments. During the transformation stage, the larval tissues break down, while imaginal discs comprising small clusters of cells, become activated, undergo proliferation and differentiation finally developing into the intricate wing structures, legs, antennae, and other adult features. These internal developments also involve the reorganization of the digestive system, restructuring of muscles, nerves, and the

Red based Jezebel

Grass yellow

Angled Castor

formation of reproductive organs. The fat bodies, serve as an energy reserve to support the metamorphosis process. A modified tracheal system is established within the pupa to facilitate respiration. Air sacs facilitate gaseous exchange supplying oxygen to the developing tissues. Finally, the adult butterfly is ready to emerge from the pupal casing, a process known as eclosion. The pupal exoskeleton splits open along a predetermined line, creating an opening through which the transformed adult gradually emerges by swallowing air and creating pressure.

4. Adult

After emergence from pupa, the wings, initially soft and crumpled, undergo expansion and hardening upon contact with air and sunlight. The adult body is prominently divided into head, thorax and abdomen.

UN

UP

Photo: Adult butterfly of red based Jezebel

Head: The butterfly head is composed of different sclerites , sense organs and the mouth parts. The fronto-clypeus occupies the anterior region of the head. The compound eyes made up of thousands of tiny lenses called ommatidia are laterally placed. They provide excellent visual acuity and a wide field of view. The ocelli are often concealed by scales or may appear to be absent.

Mouthparts: Butterflies have unique mouthpart structure adapted for feeding on nectar. This type of mouthpart is called as siphoning type formed by the maxillary part of insect mouth. It is composed of cardo, stipes, galea, lacinia and palp. The galea parts of the two elongated maxillae become apposed via hooks and muscles to form a single tube, called as proboscis. The length of proboscis normally corresponds to about 40-90% of the body length. The lumen of the proboscis forms the food canal. The proboscis is coiled up when not in use and extended to reach deep into flowers to extract nectar. A bulblike structure present in the buccopharyngeal region having narrow opening at the base of proboscis serves as pump for drawing the fluids up into the digestive system. The ability to siphon nectar with their specialized mouthparts is an essential adaptation for the survival of butterflies, as it provides them with the necessary nutrients to fuel their energy needs and has evolutionary significance. The maxillae are covered by tiny sensory hairs called sensilla, responsible for olfaction and gustation. Lepidopterans use both visual and chemoreceptive senses to locate nectar sources. The size of labial palpi shows variation.

Antennae: The antennae of butterflies are sensory structures that are used for a variety of purposes, including detecting environmental cues, identifying potential mates, and locating food sources. The flagellum is composed of many segments, typically between 10 and 20, that are connected by flexible joints. The specialized sensory structures present called sensilla, are responsible for detecting a variety of stimuli,

Mouth parts:

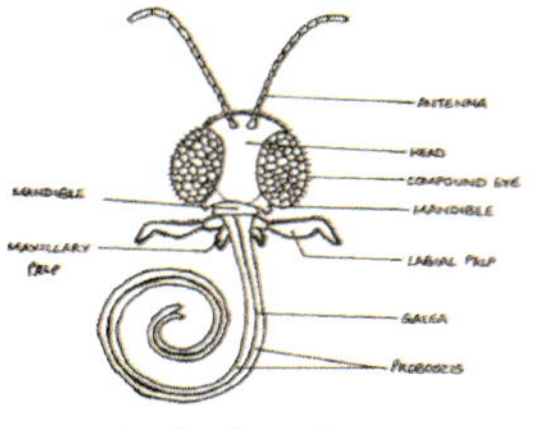

Coiled proboscis

Proboscis

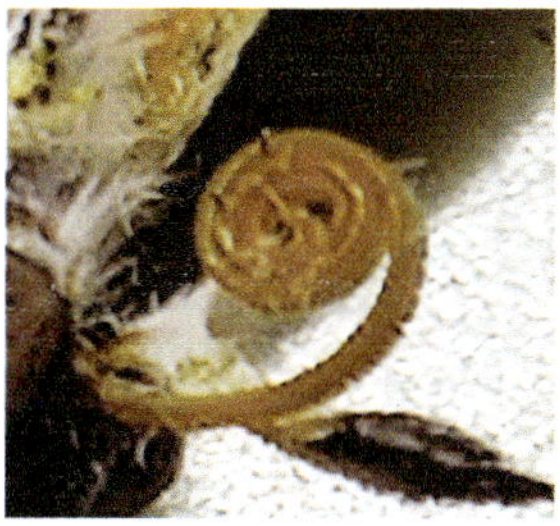

Nymphalid

Palpi

Large palp of Nymphalid

Long proboscis of Hesperid

Coiled proboscis of papilionid

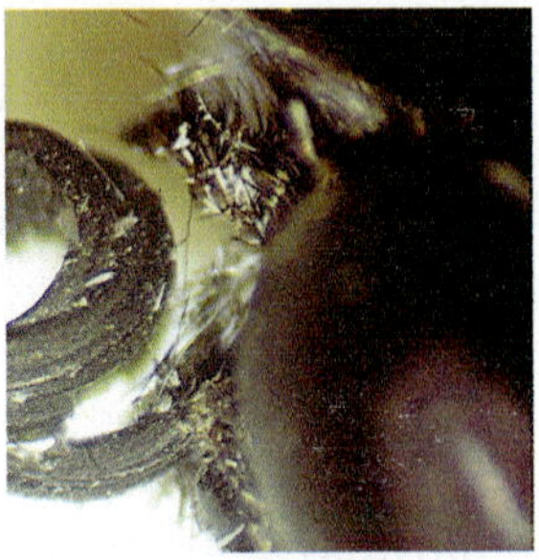

Reduced palp of Papilio

Compound Eyes:

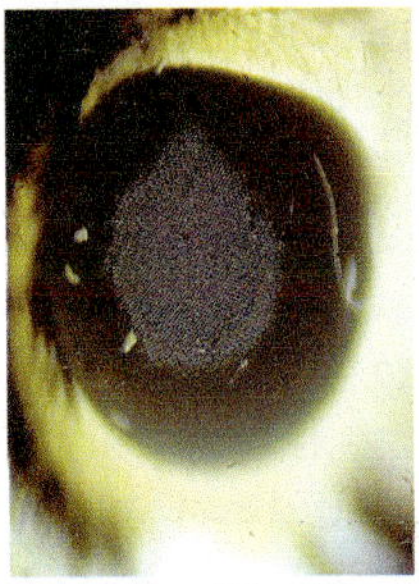

Papilionidae

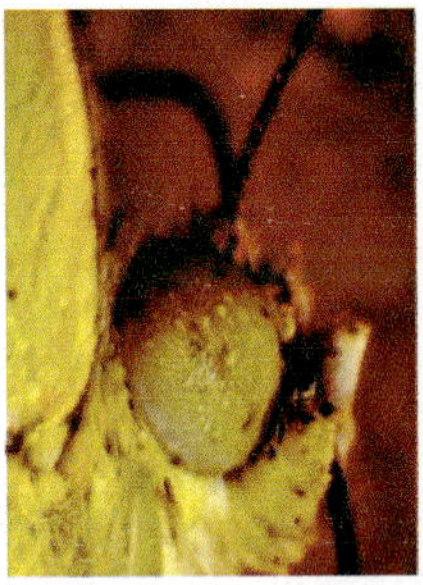

Nymphalidae

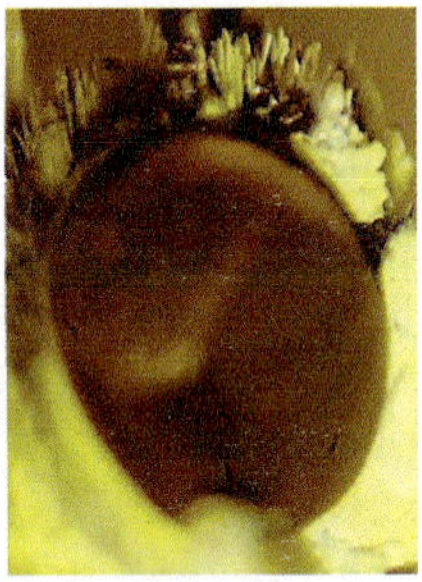

Pieridae

Antenna:

Hesperidae

Spatula like club

Bent Epiculus

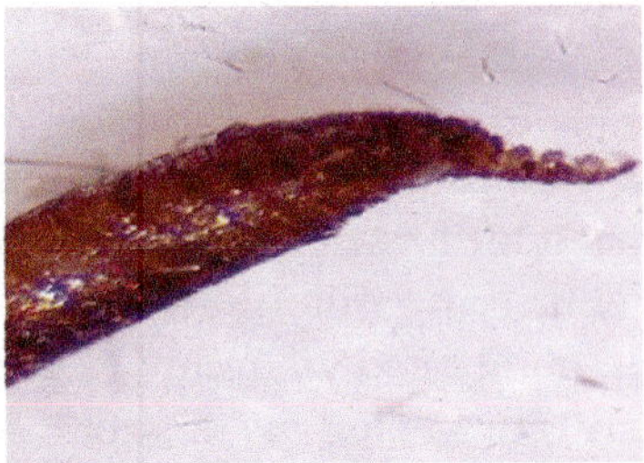

Tooth in epiculus

Papilionidae

Solid club with bent

Nymphalidae

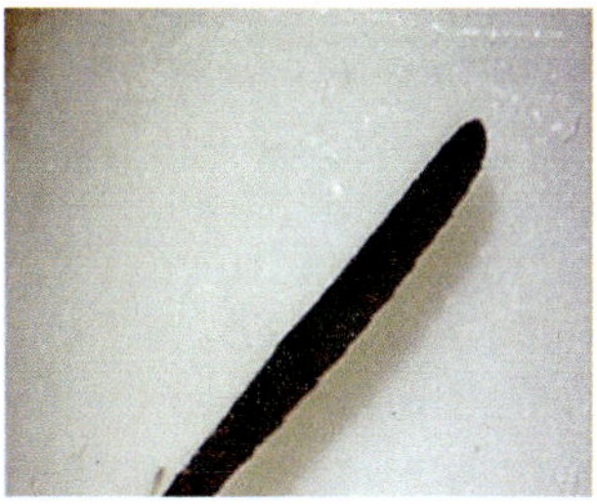

Narrow club

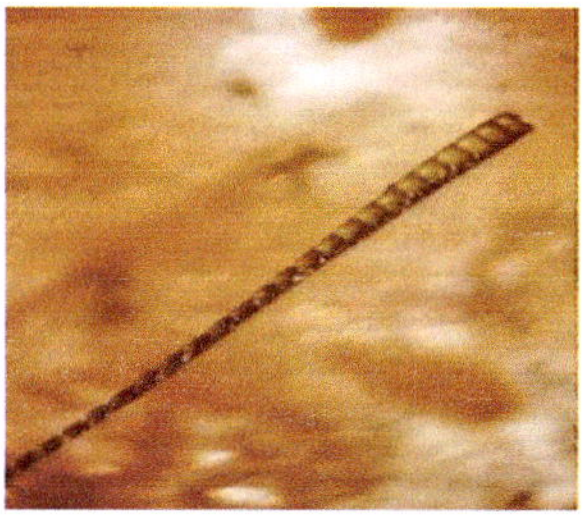

Gradual club

Pieridae

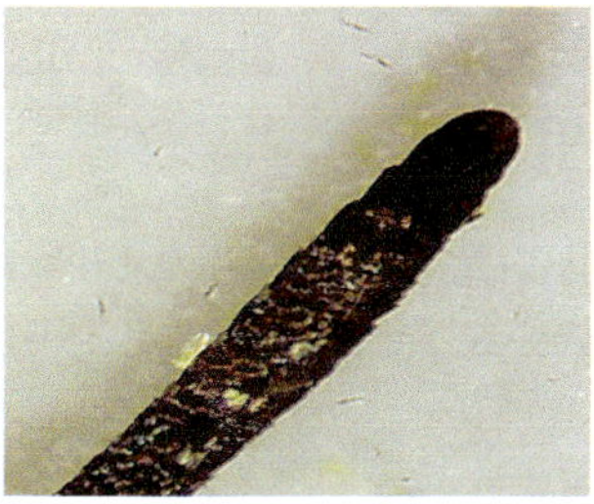

Solid club

Lycaenidae

Truncate club

including chemicals, air currents, and vibrations. The shape and size of the antennae can vary greatly between species, some have a club like structure while others have bifid hook like structure. The size of antenna may show sexual dimorphism.

Thorax: The thorax is the middle segment of the butterfly's body, to which the wings and true legs are attached. The thorax contains three segments, prothorax, mesothorax, and metathorax. The prothorax looks like a collar and may have small lateral lobes. The mesothorax is the largest segment with prominent pre-scutum, scutum and scutellum in tergum and well-developed tegulae. The metathorax is relatively small and may be greatly reduced. A meron is present in relation to the meso- and metathoracic coxae. The flight muscles attached to meso- and metathorax enable the butterflies to undertake both short and long-distance flights. The thorax is covered in a tough exoskeleton, which provides protection for the internal organs and support the muscles. The internal organs of the thorax include the respiratory system, which consists of a series of branching tubes called tracheae that deliver oxygen to the musclesthrough finer branches, the tracheoles, and the circulatory system, consisting of a dorsal blood vessel, pulsatile organs and hemolymph circulating all throughout the body. The digestive system consists of a narrow oesophagus, crop, straight, narrow ventriculus or stomach, a narrow coiled ilium, a distended colon and a short muscular rectum. Legs exhibit variations and provide good taxonomic keys.

Legs: Butterflies use their six legs for walking and clinging to surfaces. The legs have tiny spines allowing the butterfly to grip surfaces. The legs are named as pro-, meso- and metathoracic legs based on their attachment to the respective segment in the thorax. The legs are divided into the segments, coxa, trochanter, femur, tibia, and tarsus. The coxa is the base of the leg that attaches to the thorax, while the tarsus is the end of the leg that contains small segments called tarsomeres. In some butterfly families, such as the Lycaenidae, the first segment of the tarsus may be swollen in males. The skippers (Hesperiidae) and swallowtails (Papilionidae) possess a tibial epiphysis, which is absent in other butterfly families (Robbins, 1989). The legs of butterflies are generally not very mobile. The anterior legs of some butterfly families, such as the Nymphalidae, may be reduced or modified and may not be useful for walking. In Riodinidae, only in the male, the fore leg is reduced while in the female, it is functional. The tibia of the butterflies under the family Riodinidae, may have a special lamellate spur on the inner surface, which may be vestigial and might have been used for cleaning the antennae in the past. The mid and hind tibiae usually have one and two pairs of spurs, respectively, and in some cases, the posterior tibiae may have an expansible tuft of hair that functions as a scent-producing organ. The tarsi of butterfly legs are typically 5-segmented, with the first segment being the longest. The claws of the Pieridae may be cleft or bifid, and in some male Lycaenidae, one or both claws may be absent.

Scales: Butterfly scales are modified hairs that cover the wings and body . They are flattened, pigmented, and often modified setae (hairs). They are considered to be homologous with insect hairs, but are specialized for various functions such as insulation, camouflage, and mating display. The scales are arranged in overlapping rows, and they interlock with each other like shingles on a roof. Butterfly scales are made up of chitin, a polymer of n-acetylglucosamine that contributes towards sclerotization of the cuticle through cross linkage with proteins. The color of the scales is due to pigments that are either deposited within the scale or on the surface of the scale. The shape and size of the scales are highly variable between different butterfly species, and can range from round or oval to elongated and narrow. Underneath the scales, the wings are composed of a thin, transparent bi-layered membrane called the wing membrane equipped with veins. The wing veins are hollow, tubular, sclerotized structures composed of chitin and provide structural support to the wings. They contain tracheae and hemolymph vessels that transport oxygen and nutrients throughout the wings. The nervous supply to the wing is done through the nerves running inside the lumen of the veins. The arrangement of the veins is unique to each butterfly species and is elaborately used in taxonomic identification.

Membranous wings: Butterflies have two pairs of typically broad and flat membranous wings, covered in scales used for flight. They are attached to the mesothorax and metathorax through axillary sclerites. The veins include a series of longitudinal veins, costa, subcosta, radius, media, cubitus and anal. In insects, there are four axillary sclerites. The first axillary lies at the base of subcosta. It articulates with the anterior notal wing process. The second axillary lies at the base of radius and articulates anteriorly with the first axillary and the pleural wing process posteriorly. The third axillary lies at the base of anal vein and articulates with posterior notal process. These axillary sclerites give flexibility to the wings which is a land mark in insect's evolutionary history. The wing scales give the butterflies their characteristic coloration and patterns. The forewings are usually larger than the hindwings. The wing membrane is divided into the basal axillary region and membranous alar region. The alar region is further divided into remigium posteriorly lined by vannal fold, vannus posteriorly lined by jugal fold and the posterior most lobe like jugum. The borders of the alar region is named as anterior or costal margin, anterior angle or apex, apical margin or termen, posterior angle or tornus and posterior margin or dorsum. The cross veins exhibit variation in position. In wings of almost all the butterflies they form a closed discal cell which varies in shape. The hindwings are usually smaller and less complex in venation than the forewings. For the purpose of identification of butterflies, the veins are nomenclatured as 1a, 1b (anal veins), 2-12 from the dorsum. Butterflies exhibit amplexiform type

Wing venation patterns in different families of butterflies:

Family 1: Hesperiidae

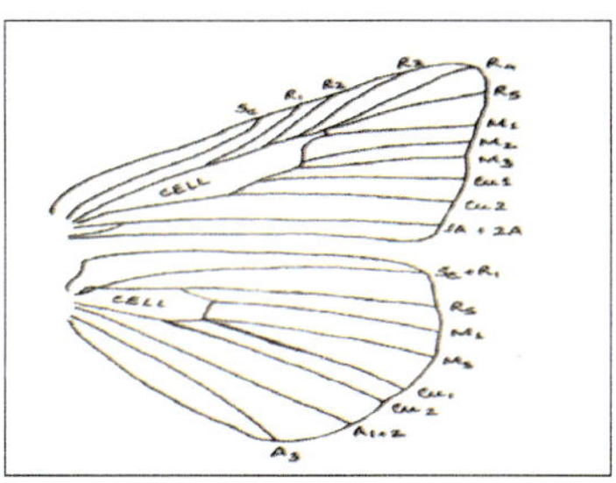

Venation of Rice swift, *Borbo*

Family 2: Papilionidae

Venation of *Papilio polytes*

Venation of Great mormon, *Papilio memnon*

Family3: Nymphalidae

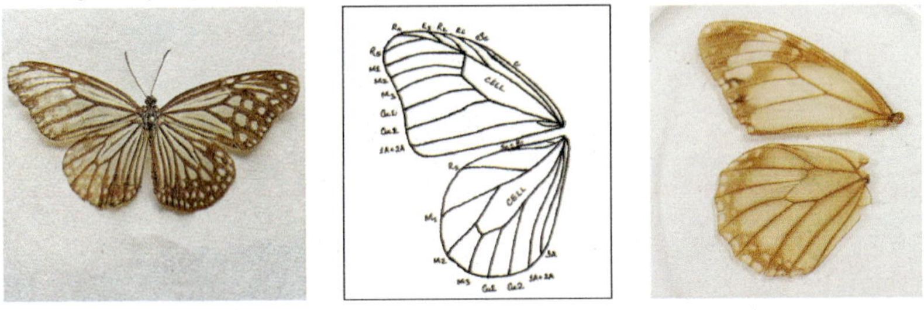

Venation of Glassy tiger, *Parantica agleas* (Left wing descaled)

Striped tiger

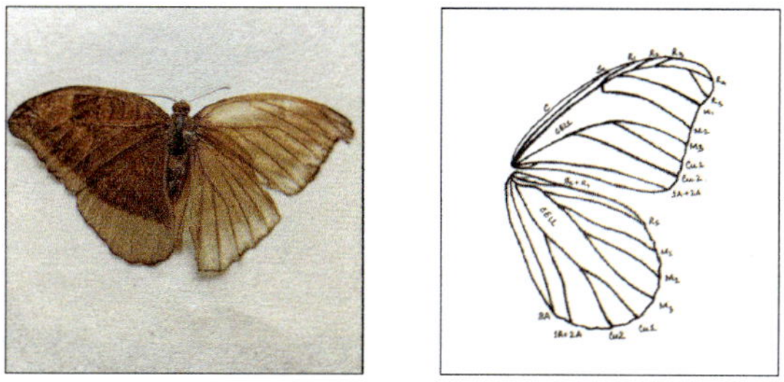

Venation of Himalayan grey count, *Tanaecia lepidea*

Family 4: Pieridae

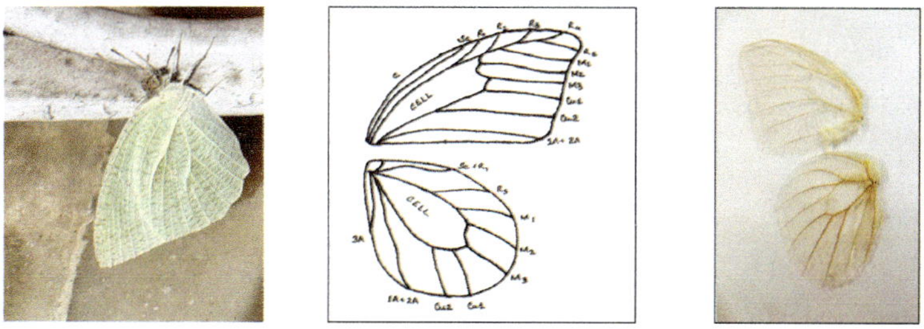

Venation of Mottled emigrant, *Catopsila* spp.

Venation of Indian cabbage white, *Pieris canidia*

Family 5: Lycanidae

Venation of Indian Grass blue, *Zizeeria*

Wing scales and patterns

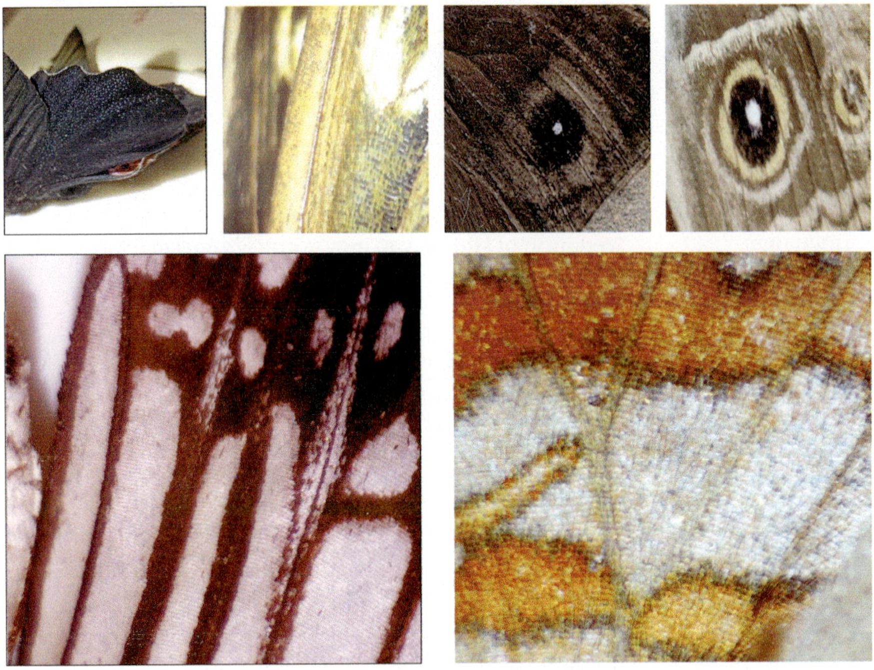

Leg morphology in different families of butterflies

Nymphalidae

Fore leg *Mid leg*

Papilionidae

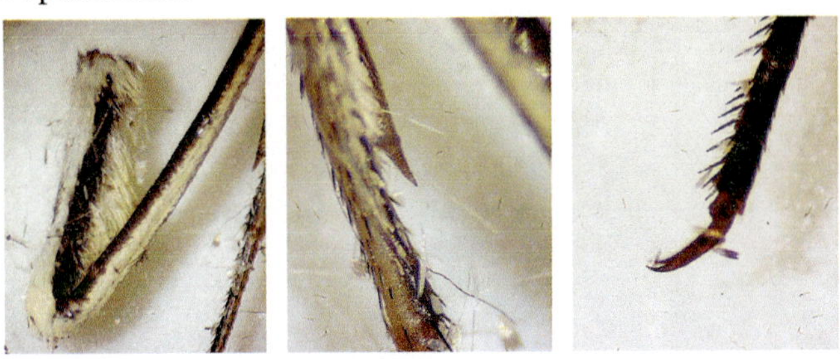

Fore leg

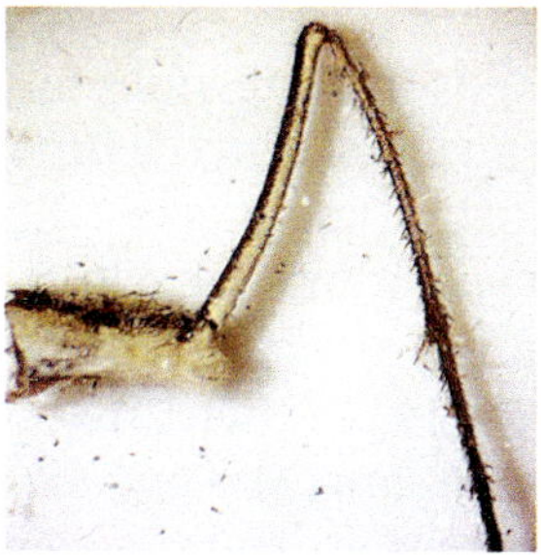

Mid leg *Hind leg*

Pieridae

Fore leg

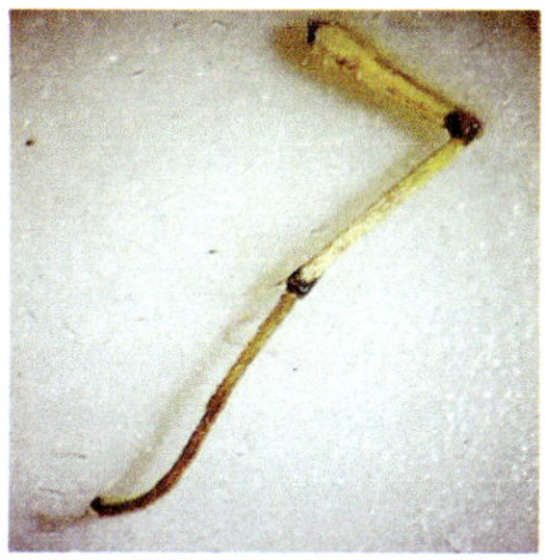

Mid leg

Hind leg

Lycaenidae

Fore leg

Mid leg

Hind leg

Hesperidae

Fore leg

Mid leg

Hind leg

of coupling apparatus in which some portions of the forewing and hind wing overlap and the power stroke of the forewing pushes down the hind wing in unison so that the two wings move together. In some species, the wings have specialized structures such as the patagia or tegulae, which are associated with the thorax and serve to streamline the body during flight. Development of the capability for flight is another landmark in the evolutionary history of insects. The wings of butterflies are used not only for flight, but also for communication, thermoregulation, and defense against predators.

Genitalia

The insect reproductive system is anatomically divided into external and internal genitalia which interact functionally. In lepidoptera the male genitalia has been used elaborately as important morpho-taxonomic key in butterfly identification. However this biasness in use of male genitaia is now questioned as both male and female genitalia exhibit diverse anatomical difference in different species . Hence their structures are studied in details by evolutionary biologists to understand the selective pressure behind origin and diversification of the genital structures and traits. Butterflies exhibit sexual dimorphism in genitalia. The males have several distinct structures, primarily designed for transferring sperms to the female during mating. Females have structures arranged for copulation, fertilization, and egg-laying. The internal duct system of females carries sperm and bear separate openings for copulation and egg-laying.

The male external genitalia includes the organs associated with the abdominal segments IX and X. The nineth abdominal segment is modified to form a ring, the dorsal side of which is called, tegumen and the ventral side is called vinculum. The tegumen is connected to uncus, a modification of segment X. The tegumen is attached on posterior part to two paired structures, gnathos, a finger like process and socii, setose lobes. The posterior end of the segment is closed by a diaphragm through which the alimentary canal opens through the anal aperture. The genital duct opens through an intomittent organ called the phallus. The latter is borne by a sclerotized plate called juxta, modified from a sclerotized part of diaphragm. The phallus may contain hooks, teeth or sclerotized structures called as cornuti. They aid in the transfer of sperm and may also function in species recognition. It delivers sperm to the female for fertilization. Additionally, males have claspers, also known as valvae or harpes, located at the posterior end of the abdomen. These are modifications of appendages of segment IX and show variations. The uncus aids in grasping and stabilizing the female during mating.

On either side of the uncus lie socii a pair of elongated, finger-like appendages. They assist in positioning and guiding the female's genitalia during copulation.

The internal reproductive system of female Lepidopterans consists of paired egg producing ovaries and accessory glands . The female reproductive system also includes receptacles and ducts for receiving, transporting, and storing sperm. The genital plate is a broad, sclerotized structure that forms the external part of the female genitalia. It serves as a protective covering for the internal reproductive organs. The bursa copulatrix is a sac-like structure (Corpus bursae) lying within the abdomen. It opens through ostium in eight sternum, receives spermatophore from male and act as a temporary storage . Corpus bursae is connected by a duct (ductus bursae) to a special seminal receptacle called the spermatheca, where it is stored until fertilization occurs during egg-laying. Eggs are laid by ovipositor which are slender tube-like structure located at the posterior end of the abdomen.

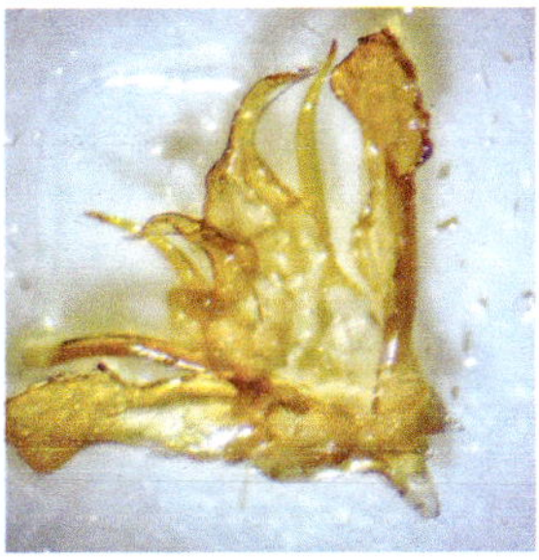

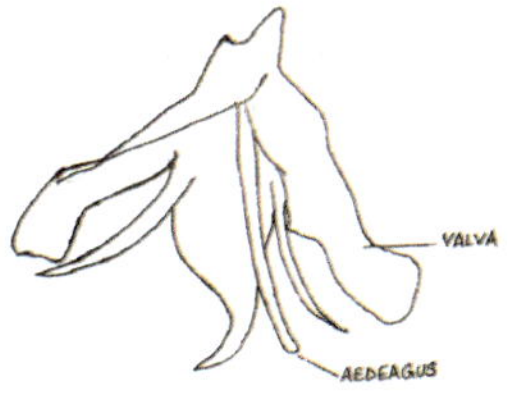

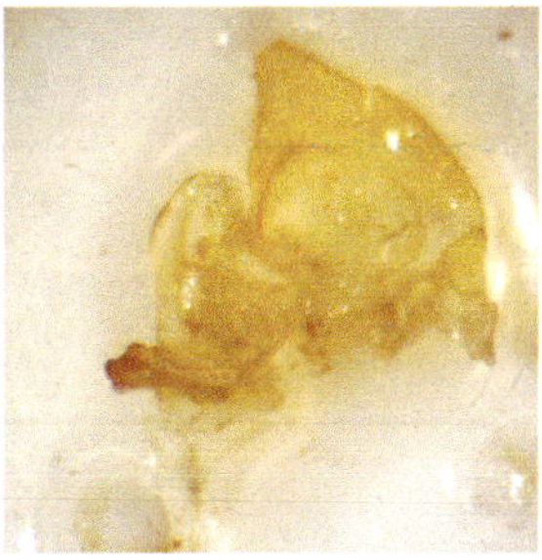

Male genitalia of *Mycalesis* (Nymphalid)

References

1. Thomson, G.(1992). Egg surface morphology of Manioline butterflies (Lepidoptera, Nymphalidae, Satyrinae). Atalanta, 23 (1/2), 195–214.
2. Yack, J. E., and Fullard, J. H. (2000). Ultrasonic hearing in nocturnal butterflies. Nature, 403 (6767),265–266.
3. Kawahara, A. Y., Storer, C., Carvalho, A. P. S., et al. (2023). A global phylogeny of butterflies reveals their evolutionary history, ancestral hosts and biogeographic origins. Nature Ecology and Evolution, 7(7),903–913. https://doi.org/10.1038/s41559-023-02041-9
4. Janz, N., and Nylin, S. (1998). Butterflies and plants: A phylogenetic study. Evolution, 52 (2), 486–502.
5. Ehrlich, P. R., and Raven, P. H.(1964). Butterflies and plants: A study in coevolution. Evolution, 18 (4), 586–608.
6. Ah-King, M., Barron, A. B., and Heberstein, M. E. (2014). Genital

evolution: Why are females still understudied? PLoS Biology, 12 (5), e10.

7. Cordero, C., and Eberhard, W. G. (2005). Interaction between sexually antagonistic selection and mate choice in the evolution of female responses to male traits. Evolutionary Ecology, 19, 111–122. https://doi.org/10.1007/s10682-004-7918-2
8. Cordero, C., and Baixeras, J. (2012). Sexual selection within the female genitalia in Lepidoptera. In Advances in Insect Physiology (Vol. 43, pp. 315–351). https://doi.org/10.1016/b978-0-12-394393-9.00007-6
9. Pšenička, J., Bek, J., Frýda, J., Žárský, V., Uhlířová, M., and Štorch, P. (2021). Dynamics of Silurian plants as response to climate changes. Life, 11 (9), 906. https://doi.org/10.3390/life11090906
10. Richard, O.W and Davies, R.G (1977) Imm's General Textbook of Entomology. Chapman and Hall, Newyork.

CHAPTER III

Guide to identification of butterflies of Assam

Class: Insecta
Order: Lepidoptera
Suborder Ditrysia
Superfamily: Papilionoidea

Common Mormon, *Papilio polytes*

Dark Glass Blue, *Tirumala septentrionis*

Family: Hesperidae

Dipsikha Bora and Amrita Mech

The biological cycle of lifeforms transfer nutrients into soil, a great contribution for regeneration of nature. Mitigation of the so called environmental crisis of imbalanced ecosystem needs maintenance of efficient nutrient cycling. Hesperids belonging to the family Hesperiidae are inconspicuous butterflies that feed on both mono- and dicotyledonous plants as larvae and take nectar from a wide variety of flowers, many even preferring those with long corolla tubes. The larvae remain in cells made of leaves and pupate within these cells. Some like the owlets are crepuscular whereas the others are diurnal . Many like the swifts and flats come out to sunbathe and can be easily found during the day time. While the leaves fall on ground due to feeding or disease or senescence , they are acted upon by microorganisms and the plant nutrients go back to soil. The same is true for the faeces and the dead bodies of the insect larvae. The hesperids are found all over the globe with 3500 identified species and many are harboured by grasses which are critical components in nutrient cycling of the biological systems. Therefore, although inconspicuous and less researched , the Hesperiids can be said to be important players in ecosystem functions and nutrient cycling. They serve as indicators of a healthy ecosystem. In Greek mythology, Hesperids are called the daughters of the evening , the protectors of mythical garden growing golden apple. The hesperid diversity is observed to be more in areas of north east India having extensive area of betelnuts and banana trees.

The Hesperid butterflies are also known as 'skippers' because of the habit of fast erratic darting, skipping pattern of movement facilitated by their strong legs, conspicuous thorax and small wings. They are characterized by their unique kind of antenna bearing a club produced into a hook like structure called as epicule. They have stout bodies as compared to other families of butterflies. Because the bodies are covered with long hairs and profuse scaling, they appear moth-like and amateurs confuse them with moths. The skippers are mostly dull in colour but some with colourful scales giving specific patterns which may be used in taxonomic identification. Other characters used for taxonomic identification are size, colour, shape of epicule on the antenna, nudum, length of antenna, wing shape, wing venation pattern, wing coupling apparatus, leg structure, presence of hair brush and structure of external genitalia. During resting or basking, some of them keep their hind wings flat and the forewings at about forty-five degree angle up which can be used as ready identification key for categorization into Hesperiidae family. Hesperiinae is identified by the positioning of vein 5, the presence or absence of a constriction at the joint of antennal club and epicule, antennal club and epicule structure, palpi structure (porrect or not) and size of wing and wing cell. Pyrginae are identified instantly by flatly placed wings when in repose. The other characters considered are positioning of vein 5, comparative length of abdomen and dorsum, wing cell length, palpi structure. However, some ground breaking molecular studies have also revealed new phylogenetic relationships and intriguing questions regarding placement of some genera based on morphological treatments .

Out of 4113 species of Hesperiids so far identified in the world, 341 are recorded from India. In this book we are referring to only 25 very common hesperiids found in and around human habitations in Assam which represent five subfamilies, Hesperinae, Teractocerinae, Heteropterinae, Coeliadinae and Pyrginae.

Subfamily 1: Hesperiinae

1. Dark Branded Swift, *Caltoris brunnea caere* de Nicéville

Subfamily: Hesperiinae

Genus: *Caltoris*

Species: *brunnea*

Subspecies: *caere*

Distinguishing characters: Dark brown with clothing of greenish hair on thorax and wingbase. In Hesperiinae, upper fore spots at 2&4 are separate from spots in 6-8. In *Caltoris*, under hind unmarked . The mid tibia is not spiny, bears a pair of spur. Antenna brown with flattened club, and epicule not at right angle. In this species *brunnea caere*, upper fore with two apical yellowish spots in 6-7 and three in decreasing order of size in 1b,2,4. Forewings pointed and the two cell spots upperfore are separate. Cilia brownish, white in tornus . Rest with wings closed. Wing span: 30mm.

Global Distribution: India, Bangladesh, Myanmar.

Recorded Localities and season: Dibrugarh, Post Monsoon

Land use and land cover pattern: Basks in shrubs under the sun. Found in open garden.

Forage plant: Asteraceae, Fabaceae

2. **Veined Palmer, *Hidari bhawani*** de Nicéville

   Subfamily: Hesperiinae

   Genus: *Hidari*

   Species: *bhawani*

Distinguishing characters: Upper dark brown, under pale brown. Tip of antenna white. In *Hidari,* the vein 5 in fore is decurved at origin. Upper fore with hyaline spots. Under hind with inconspicuous spots and with red striations. In *bhawani,* costa and apex of fore dark brown. Fore with only one apical spot . UNH with white centered dark brown spots making a half circle. One prominent spot at the base of v7 underhind. Distinct red streak v8 and toward dorsum underhind. Cilia chequered black and brown. Club of antenna white and without prominent hook. Wing span: 45 mm. I could not match it with any type specimen. But it was recorded in Lakhimpur District of Assam as mentioned in Evans,1934.

Global Distribution: India, Bangladesh, Myanmar.

Recorded Localities and season: Charaideo, November

Land use and land cover pattern: Found in open garden, attracted to light

Forage plant: Not recorded

3. **Rice Swift, *Borbo impar* Mabille**

   Subfamily: Hesperiinae

   Genus: *Borbo*

   Species: *impar*

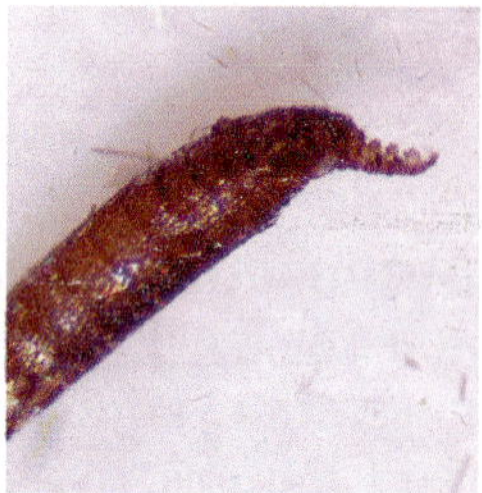

Distinguishing characters: Dark brown with transparant marks. In Hesperiinae, upper fore spots at 2-4 are separate from spots 6-8. In *Borbo*, unh with no cell spot and mid tibia not spined, bears two pairs of spurs. In *impar,* fore has transparent spot 1b-4 and apical 6-8 . The spot in 1b is excavated square. Upperhind plain. A small spot in cell upperfore. Underhind has transparent squared spot 2-4. Upper with two ochreous hairbrush attached to both sides of thorax and base . Antennal epiculus at obtuse angle and with hooks. Wing span: 40mm

Global Distribution: India (Assam, Sikkim, Bengal), Andaman, Hainan, Borneo

Recorded Localities and season: Dibrugarh, May

Land use and land cover pattern: Found in open garden.

Forage plant: Asteraceae, Poaceae(Grasses), Apocynaceae (*Catharanthus*)

4. Rice Swift, *Borbo cinnara* Wallace

Subfamily: Hesperiinae

Genus: *Borbo*

Species: *cinnara*

Distinguishing characters: Upper greyish brown with greenish scale or hairs on thorax. Under pale brown. Forewing pointed. In *Borbo* mid tibia unspined and no cell spot in hind. In *cinnara*, upperfore has a cell spot, the spot at 2 quadrate and the spots from space 2 to 4 decrease in size. The upperfore apex bears spot at 6-8. Upperhind spot at 2,3,4 and the same are observed in underhind. The upperhind spot in 4 is said to be rarely seen. Wing span: 30-36mm

Global Distribution: India, Bangladesh, Myanmar , Bhutan, Bangladesh

Recorded Localities and season: Sivasagar, June

Land use and land cover pattern: Basks in shrubs under the sun. Found in scrub.

Forage plant: Poaceae

5. Small Branded Swift, *Pelopidas mathias* Fabricius

Subfamily: Hesperiinae

Genus: *Pelopidas*

Species: *mathias*

Distinguishing characters: Body and wing golden brown, with greyish hair on thorax. Under hind is greyish brown toward dorsum. Forewing pointed. Underhind small white spots , one in cell and others in spaces 2,3,4,6 .Mid tibia without spur. Antenna with epicule curved. Wingspan: 32-38mm..

Global Distribution: India,Java, Bhutan, Nepal, Sri Lanka, Bangladesh, Myanmar

Recorded Localities and season: Buxa, Dibrugarh, October

Land use and land cover pattern: Shrubs and grasses.

Forage plant: *Lantana*

6. Great Swift, *Pelopidas assamensis* de Nicéville

Subfamily: Hesperiinae

Genus: *Pelopidas*

Species: *assamensis*

Distinguishing characters: Very dark greyish brown. In *Pelopidas*, underhind bears a cell spot and mid tibiae with prominent spur. In *assamensis*, fore with two transparent cell spots contiguous. Fore with spots 1b,2,4 and three in 6-8. UNH covered with dark ochreous grey hairs, spots in vein space and one in cell very small. Club of antenna very dark brown. Apical marks are squared and apically excavated. Spot in 4 elongate. Antennal club narrow pale below club. Legally protected under Schedule IV of IWPA, 1972 . Wing span: 46mm

Global Distribution: India, Bangladesh, Myanmar , Bhutan, Bangladesh

Recorded Localities and season: Sivasagar, June

Land use and land cover pattern: Basks in shrubs under the sun. Found in open garden.

Forage plant: Poaceae, Asteraceae

7. Conjoined Swift, *Pelopidas conjuncta* Herrich-Schaffer

Sub family: Hesperiinae

Genus: *Pelopidas*

Species: *conjuncta*

Distinguishing characters: Light brown with pale yellow spots and green suffusion at base. Upperfore with two separate cell spots and variable discal spots. UPH is plain. UNH with three spots at 2,3 and 4, the spots 2,3 in line.

Global Distribution: India, Nepal, Bhutan, Bangladesh, Myanmar., Sri Lanka

Recorded Localities and season: Dibrugarh. Season : Post Monsoon (August-September)

Land use and land cover pattern: Basks under sun in shrubs. Found in open garden .

Forage plant: *Catharanthus*

8. Lesser Rice Swift, *Borbo bevani* Moore

Sub family: Hesperiinae

Genus: *Borbo*

Species: *bevani*

Distinguishing characters: Brown, upper with ochreous hairs at base of thorax. In *Borbo bevani*, upper fore with small spot at 1b. Hindwing termen rounded.Underhind white spot from 2-5, 4 being indistinct. Wingspan:32-36

Global Distribution: India, Burma, Myanmar, Philippines

Recorded Localities and season : Dibrugarh. Season : Post Monsoon (August-September)

Land use and land cover pattern: Basks under sun in shrubs. Found in open garden .

Forage plant: *Catharanthus* or flowers with long corolla

9. Himalayan Small Palm Bob *Suastus minuta aditia* Evans

Subfamily: Hesperiinae

Genus: *Suastus*

Species: *minuta*

Subspecies: *aditia*

Distinguishing characters: Body and wing upper greyish brown tinge with white spots. In *Suastas,* under hind brownish with black non-transparent spots, in a decreasing row from 1c to3, 6 and upper edge. In *minuta* , upper fore with one mark cell- end . Dot like three transparent spots towards apex v6-8 and square shaped transparent marks v1c, 3,4. A black mark in cell of UNH. Antenna with a curved club. Wing span:31mm

Global Distribution: India, Sri Lanka, Bangladesh, Myanmar.

Recorded Localities and season: Dibrugarh, Spring

Land use and land cover pattern: Bushes, shrubs

Forage plant: Arecaceae

10. Indian Palm Bob, *Suastus gremius* Fabricius

Subfamily: Hesperiinae

Genus: *Suastus*

Species: *gremius*

Distinguishing characters: Brown. In *Suastus,* upper spots hyaline, v2 opposite of v7. Underhind brown with pale yellow scales with distinct black spot and not transparent. In *gremius* ,upper spot in v6-8 contiguous and in line. Margin with pale yellow cilia. Rests with wings folded. Antenna with hook making obtuse angle with club. Wing span:37mm

Global Distribution: India, Bhutan

Recorded Localities and season: Dibrugarh, Monsoon, Post Monsoon (August)

Land use and land cover pattern: Shrubs and grasses.

Forage plant: Arecaceae, Seasonal flowers, Larval host plant: Palm trees

11. Common Red Eye, *Matapa aria* Moore

Subfamily: Hesperiinae

Genus: *Matapa*

Species: *aria*

Distinguishing characters: Dark brown. In *Matapa*, antenna half costa in length. Epiculus at right angle to club. Fore cell apex produced. In hind, v7 origin is well before origin v2. In *aria*, the cilia yellow in hind. Underhind ferruginous brown with basal dull brown area. Antenna club flatten and with a curve. Wing span: 40-55mm

Global Distribution: India (Assam, Arunachal Pradesh, South India, Uttaraphand, J&K), Bhutan, Nepal

Recorded Localities and season: Sivasagar, Autumn

Land use and land cover pattern: Basks in shrubs under the sun. Found in open garden, attracted towards light.

Forage plant: Solanaceae

12. Dark Velvet Bob, *Koruthaialos butleri* WM & de Nicéville

Subfamily: Hesperiinae

Genus: *Koruthaialos*

Species: *butleri*

Distinguishing characters: In *Koruthaialos,* body and wing dark velvety brown above and below and bears no hyaline spot. UNH unmarked. Antenna equal half costa, epiculus gradual. In *butleri,* upper fore discal

red band just indicated along v2 . Wing span: 38mm

Global Distribution: India (Assam, Arunachal Pradesh), Bangladesh, Bhutan.

Recorded Localities and season: Dibrugarh, Spring (March)

Land use and land cover pattern: Shady places undergrowth

Forage plant: Not listed

13. Bengal Chocolate Demon, *Ancistroides nigrita diocles* Moore

Subfamily: Hesperiinae

Genus: *Ancistroides*

Species: *nigrita*

Subspecies: *diocles*

Distinguishing characters: Dark brown with paler shade underwing towards margin. In *Ancistroides,* no hyaline marks. V4 is closer to v5 than v3. In *nigrita,* no marks above and under. Proboscis is very long. Eyes dark brown. Cilia brown. In *diocle,* under fore apex, under hind dorsum paler than the rest. Observed often resting on damp wall with wings closed. Keep their wings closed even during nectaring. Antenna club bent making an obtuse angle.Wing span: 48-60mm

Global Distribution: India, Bhutan, Bangladesh, Myanmar.

Recorded Localities and season: Dibrugarh, Post Monsoon (August)

Land use and land cover pattern: Visits flower and damp patches.

Forage plant: *Ixora, Clitoria* (Aparajita)

14. Grass Demon, *Udaspes folus* Cramer

Subfamily: Hesperiinae

Genus: *Udaspes*

Species: *folus*

Distinguishing characters: Body and wing dark brown. In *Udaspes* wings with hyaline marks. Antenna half costa in length. In *folus* , the mark in lower border of cell is cojoined with the mark of 1b. Fore with three apical white marks in v6-8. Hindwing with semi-transparent central white mark . Cilia chequered brown and white. Antennal club thin.Wing span: 40-48

Global Distribution: India, Bhutan, Sri Lanka, Nepal, Bhutan.

Recorded Localities and season: Dibrugarh, Post Monsoon (August)

Land use and land cover pattern: Shrubs and grasses.

Forage plant: Nectaring *Catharanthus*

15. Spotted Demon, *Notocrypta feisthamelii* Boisduval

Subfamily: Hesperiinae

Genus: *Notocrypta*

Species: *feisthamelii*

Distinguishing characters: Body and wing dark chocolate brown. The discal band underfore is continuous with a yellow spot in fore-costa. Wing span: 38-50

Global Distribution: India, Bhutan, Sri Lanka, Nepal, Bhutan.

Recorded Localities and season: Buxa, Dhuburi; October

Land use and land cover pattern: Shrubs and grasses, open gardens near forests

Forage plant: Nectaring *Ixora,* compositae

16. Tree Filter, *Hyarotis adrastus* Stoll

Subfamily: Hesperiinae

Genus: *Hyarotis*

Species: *adrastus*

Distinguishing characters: Under dark brown with irregular white central band. In *Hyarotis adrastus,* in underfore, the white cell mark extends to costa and the mark at 1b is diffused. Under fore apex with three white marks in v6-8. Antenna with epicule. Move over tree canopies. Protected under Schedule IV of IWPA, 1972. Wing span: 38-48mm

Global Distribution: India, Bhutan, Nepal, Bangladesh, Myanmar.

Recorded Localities and season: Kaziranga, Post Monsoon (July-August)

Land use and land cover pattern: Shrubs and grasses.

Forage plant: *Clerodendrum* (Verbenaceae)

Subfamily 2: Taractrocerini

17. Lydia Dart, *Potanthus lydia lydia* Evans

Subfamily: Taractrocerini

Genus: *Potanthus*

Species: *lydia*

Subspecies: *Lydia (fraseri if spot at 6 absent)*

Distinguishing characters: Upper dark brown with olive green hair and scales in thorax. In Teractrocerini, upper fore spots from 1b to 8 separated by dark brown veins, but not in line, in 4&5 shifted out. Mid tibiae smooth. Spots in cell conjoined. UN fore with marks similarly with UP as they are transparent. In *Potanthus,* antenna equals half costa and epicule is hooked. Upperfore marks 4&5 smaller than the epical marks. In, *lydia* under hind with five tawny elongate marks from space2 to 6, and 5 & 6 cojoined. Antennal club flat with a pointed epiclulus. Brown cilia along margin are bright orange in tornus. Wing span: 46mm

Global Distribution: India (Meghalaya, Nagaland, Assam), Bangladesh, Myanmar Java

Recorded Localities and season: Urban garden,Dibrugarh, riverside forest treks, Monsoon

Land use and land cover pattern: Basks in shrubs under the sun. Found in open garden.

Forage plant: Not recorded

18. Common Dartlet, *Oriens gola pseudolus* Mabille

Subfamily: Taractrocerini

Genus: *Oriens*

Species: *gola*

Subspecies: *pseudolus*

Distinguishing characters: Body and wing dark brown. Wings with tawny yellow marks. In Teractrocerini, upper fore spots from 1b to 8 with continuous ,but not in line, in 4&5 shifted out. Mid tibia smooth. In *Oriens*, upperfore with tawny yellow almost a central band made up of broad marks , spot in 1b is centrally placed. Antenna greater than half wing. Spot in cell is conjoined to discal band. In *gola pseudolus*, upperfore with cell spot and tawny mark near basal costa . Spot in 5 is connected to apical marks in 6-8. Upper hind with tawny yellow central band, Under hind similar as upper. During sitting, keep the hind pair open and fore pair closed. Rapidly fly for moving from one flower to another for nectaring. Egg laying observed in *Bougaivillea*. Antenna club with short epicule. Wing span: 30mm

Global Distribution: India,Java, Bhutan, Nepal, Sri Lanka, Pakistan, Bangladesh, Myanmar.

Recorded Localities and season: Dibrugarh, Post Monsoon (August)

Land use and land cover pattern: Shrubs and grasses.

Forage plant: Clerodendrum (Verbenaceae)

19. Palm Dart, *Telicota* sp. Moore

Subfamily: Taractrocerini

Genus: *Telicota*

Distinguishing characters: Under tawny yellow with dark brown central and postdiscal marks. Marginal black border. Tibia smooth. Antenna with black and white marks. In *Telicota,* antennae not longer than half costa. Shaft thin. Hind cell equals half of wing. Fore wings produced. Veins yellow. One pair of tibial spur missing in hind tibia. Spur in mid tibia absent. Nectaring in Solanaceae. Wingspan: 28mm..

[The insect is compared with specimens of NCBS website]

Global Distribution: India,Java, Bhutan, Nepal, Sri Lanka, Pakistan, Bangladesh, Myanmar.

Recorded Localities and season: Urban garden, Dibrugarh, Post Monsoon (August)

Land use and land cover pattern: Shrubs and grasses.

Forage plant: *Clerodendrum (*Verbenaceae), *Catharanthus*

Subfamily3: Heteropterinae

20. Forest Hopper, *Astictopterus jama* C & R Felder

Subfamily: Heteropterinae

Genus: *Astictopterus*

Species: *Jama*

Distinguishing characters: DC: Dark brown. Unmarked except for some apical white spots. Fore wing produced. Indistinct band or reddish scaling may be found in UNH. Wingspan: 42-50mm

Global Distribution: India, China, Philippines

Recorded Localities and season: Fringe areas of forest, Dibrugarh, August

Land use and land cover pattern: Shrubs and grasses.

Forage plants: Not recorded

Subfamily 4: Coeliadinae

21. Common Banded Awl, *Hasora chromus* Cramer

Subfamily: Coeliadinae

Genus: *Hasora*

Species: *chromus*

Distinguishing characters: Brown wings and body. Thorax strong. In Coeliadinae, eyes large, dark brown. In *Hasora*, fore wings highly produced. Hindwing round. Head and thorax clothed with green hairs. In *chromus*, underhind, base greenish, a white band outwardly mildly diffused. Black mark at the lower end of the white band. Male upper without any mark. Observed nectaring at the onset of dusk. Flies away very fast on disturbance. Wing span: 45-50mm

Global Distribution: India (Assam, Sikkim), China

Recorded Localities and season: Dibrugarh, Autumn (Oct-Nov)

Land use and land cover pattern: Shrubs and grasses.

Forage plant: Marigold (*Tagetes*)

22. Brown Awl, *Badamia* sp. Moore

Subfamily: Coeliadinae

Genus: *Badamia*

Distinguishing characters: Forewings highly produced and hind wing equals almost two third the length of forewing. Hindwing tornus rounded. Eyes coffee brown. Body and wings coffee brown, unmarked except for some hyaline spots in fore and dark ferruginous sheen underhind. Mid

tibia smooth and with one pair of spur. Hind tibia with spines and two pairs of spurs. Underhind with purplish chocolate marks, Underfore paler. Termen with yellow cilia. Fore with yellow subapical spots which are not found in *Badamia exclamationis*. Femur with bushy hairs, tibia with fringes. The identification of the specimen needs more study and requires matching with a type specimen. In absence of a type specimen, it may be a new report.

Global distribution: India, Nepal, Bhutan

Recorded Localities and season: Dibrugarh, Post Monsoon (August), Winter (Feb)

Land use and land cover pattern: Shrubs and grasses, open garden

Forage plant : Nectaring on *Catharanthus*

Subfamily5: Pyrginae

23. Fulvous Pied Flat, *Pseudocoladenia dan dan* Fabricius

Subfamily: Pyrginae

Genus: *Pseudocoladenia*

Species: *dan dan*

Distinguishing characters: Wings upper and body golden orange with bright rufous scales scattered. In *Pseudocoladenia*, upper fore with hyaline marks near apex and discal region. Upper hind with dark marks. Antenna with epiculus. In *dan dan*, female, the mark in fore cell outwardly excavated and apical spots are not in line. Antenna with sharp hook. Wing span: 40mm

Global Distribution: India(Assam, South India), Nepal, Bhutan, Bangladesh, Myanmar.

Recorded Localities and season: Dibrugarh, Dibrugarh Univ. campus, Post Monsoon (Nov)

Land use and land cover pattern: Basks under sun in shrubs. Found in open garden near forest

Forage plant: Not listed

24. Common Spotted Flat, *Celaenorrhinus leucocera* Kollar

Subfamily: Pyrginae

Genus: *Celaenorrhinus*

Species: *leucocera*

Distinguishing characters: Dark brown. In *Celaenorrhinus* upper fore apex bear small white spots. Semi transparent large white patches from costa to dorsum in upperfore. In *leucocera,* upperhind with submarginal small yellow spots and cell end with yellow line. Yellowish hair at base.

Legally protected under Schedule II of IWPA, 1972. Wing span: 45-55mm

Global Distribution: India (Assam, South India, Jammu Kashmir), Nepal, Bhutan, Bangladesh, Myanmar.

Recorded Localities and season: Sivasagar, March

Land use and land cover pattern: Basks under sun in shrubs. Found in open garden .

Forage plant: Not listed

25. Common Snow Flat, *Tagiades japetus* Stoll

Sub family:Pyrginae

Genus: *Tagiades*

Species: *japetus*

Distinguishing characters: Upper brown. Upperfore with three apical white spots and two transparant larger spots in space 2 and 3. Discal faint dark brown spots upperhind not in line. Protected under Schedule II of IWPA,1972Wingspan: 40-45

Global distribution: India, Bangladesh, Myanmar.

Recorded Localities and season: Dehing Patkai. Season : May

Land use and land cover pattern: Basks under sun in shrubs. Found in open garden .

Forage plant: Marigold (*Tagetes*)

Family: Papilionidae

Dipsikha Bora, Phalgun Chetia, Pooja Nag

The commonly known swallowtail butterflies belonging to family Papilionidae constitute the most magnificent group of butterflies. The name swallowtail is given due to the presence of a tail in the hindwing which is however lacking in some groups of Papilionids. This is the smallest family of butterfly with fewer than a thousand species described in the world. They are found in all possible environmental habitats from high altitudes of Himalayas to lower valleys, dense canopy forest to degraded forest, scrubs and gardens.

The Papilionids have comparatively longer bodies, long legs, long proboscis, small palpi. Body colour may vary and accordingly are grouped into black bodied, red bodied swallowtails. Wing scale colours widely vary, organized into spots, streaks, ocelli and brand. They are identified mostly by positional origin of veins, cell pattern, scale designs on wings. Males may have different colour pattern than females. The same species may have different forms. Eggs are laid singly. Larvae may be greenish in colour, some exhibiting mimicry of bird droppings and displaying sluggish behaviour. Larvae possess a forked protrusible glandular structure between the head and thorax called as osmeterium, which is coloured and may serve for aposematic warning for colour or smell. The arms of the fork at base possess two exocrine glands and when disturbed they release offensive chemical. The chemicals are identified in a specialist Papilionid of Troidinae as mono- and sequiterpenoids which are repelling to predatory ants.

Like all other insects, Papilioniids have temperature thresholds and either undergo diapause or migrate for adaptation to seasonal change of temperature and non-availability of hostplants. They may be observed to be engaged in nectaring very often in plants of compositae family where with less expenditure of energy they can have access to large number of nectar sources. Several Papilionids are frugivorous and some others visit cattle droppings. They appear very restless with frequent fluttering of wing during nectaring. In forest tracks, ~~a~~ large number of different Papilionids may be seen to come out for mud puddling. We may observe congregation of the same species or many different species. For mating they are observed to fly high and engage in a dance while trying to entangle. About 107 species have been detected in India and we are reporting 24 species including several forms of *Papilio* in this book. Out of three subfamilies, Baroniinae, Parnassinae and Papilioninae, the butterflies reported in the book represent only Papilioninae and three tribes. A good number of Papilionids can be recorded from the forests of Assam.

Tribe1:Leptocircinae (Dragontail, Swordtail, Jay, Zebra)

1. White Dragontail, *Lamproptera curius* Fabricius.

Subfamily: Papilioninae

Genus: *Lamproptera*

Species: *curius*

Distinguishing Characters: Body black. One third of the wing is a black bordered triangle with black stipes. In *Lamproptera,* hind wing produced into white tipped long tail (23-38mm). Antennae longer than cell. In *curius,* the fore and hind upper bear white bands. Hindwing tornus has white marginal border. Wingspan: 41-56 mm. uncommon .

Global Distribution: Assam, Arunachal Pradesh, Nagaland, Myanmar.

Recorded Localities: Near Panbari reserve forest, Kaziranga

Land use land cover pattern: Open sunlight areas in forest

Forage plant: Not recorded

2. Five Bar Swordtail, *Graphium antiphates nebulosus* Butler

Synonym: *Graphium antiphates* Cramer, 1775

Subfamily: Papilioninae

Genus: *Graphium*

Species: *antiphates*

Sub species: *nebulosus*

Distinguishing Characters: In *Graphium,* hind wing produced into a black sword like tail with white edge. Pale greenish with black stripes. The hindwings produced into swords. In *antiphates,* bar2 from the base of v1. Underhind with central row of pale yellow marks. Upper and underfore cell with five bars. Hind with brownish black tornal area. Fast flier. Wing span:80-95mm

Global Distribution: India, Sikkim, Nepal, Bhutan, Srilanka, Bangladesh

Recorded Localities and season: Kaziranga; Summer

Land use land cover pattern: Woodland, forest edge, near water source

Forage plant: Forest flowers, genus not recorded

3. Oriental Tailed Green Jay, *Graphium agamemnon* Linnaeus

Subfamily: Papilioninae

Genus: *Graphium*

Species: *agamemnon*

Distinguishing Characters: Black with bright green spots and streaks on both FW and HW. In *Graphium,* hind tail short and black . In *agamemnon,* upper fore with cell spots in double row. UN brown with less prominent green spots. Orange bordered lunule under hind. Fore wing narrow

and elongated towards apex. Fast fliers. Rest by opening both wings. Wingspan: 80-90mm

Global Distribution: India, Sikkim, Arunachal Pradesh, Bhutan

Recorded Localities and season: Dibrugarh, Tezpur, Guwahati, Kokrajhar; Monsoon

Land use land cover pattern: Found mostly on upper canopies in forests, bush levels in parks

Forage plant: *Ixora* spp., *Lantana camara, Hydrangea*

4. Himalayan Common Jay, *Graphium doson axion* Felder and Felder

Synonym: *Graphium doson axionides*

Subfamily: Papilioninae

Genus: *Graphium*

Species: *doson*

Subspecies: *axion*

Distinguishing Characters: Hind wing tail absent. In *Graphium,* hind wing tail may be absent. Black with pale blue central band and irregular patches and marginal row of blue spots. UNH dorsum with red band followed by three postdiscal red marks in v1-3. In *doson axion,* under hind with costal red mark. Body with black and white stripes. Wing span: 70-80mm

Global Distribution: India, Sri Lanka, Nepal, Bhutan, Myanmar

Recorded Localities and season: Kokrajhar; Monsoon

Land use land cover pattern: Forest margins including damp places,

Forage plant: Forest flowers, *Ixora, Bougainvillea*

5. **Oriental Common Bluebottle, *Graphium sarpedon sarpedon* Linnaeus**

   Subfamily: Papilioninae

   Genus: *Graphium*

   Species: *sarpedon*

   Subspecies: *sarpedon*

Distinguishing Characters: Hind with short tail. UNH has black spots bounded by red placed posterior to the discal band. The hind wing termen with four submarginal blue crescents in a row. Body with pale white stripes. In *sarpedon*, under hind base with red mark. Black with central blue band in both wings tapering towards dorsum and costa. Protected under Schedule II of IWPA,1972. Wing span: 80-90mm

Global Distribution: India, Bhutan, Nepal, Bangladesh, Pakistan

Recorded Localities and season: Kokrajhar, Dibrugarh; Monsoon

Land use land cover pattern: open places near dense vegetation or forest, canopies

Forage plant: Forest flowers, *Lanatana* spp., *Cassia fistula*

6. East Himalayan Lesser Zebra, *Graphium macareus indicus* Rothschild

Subfamily: Papilioninae

Genus: *Graphium*

Species: *macareus*

Subspecies: *indicus*

Distinguishing Characters: Tailless. Black with bluish white marks. Upper fore cell with four transverse white stripes, two spots endcell and two elongated stripes beyond.Upperfore with costal row of white spots. Under brownish, marking light. Upperhind marginal row of white crescents in1-4 . In *Macareus indicus,* white stripes are broad. Hind tornus with distinct orange marks. Wing span: 80-90mm

Global Distribution: India, Bhutan, Nepal, Bangladesh, Myanmar

Recorded Localities and season: Kokrajhar, Udalguri; Monsoon

Land use land cover pattern: open places near dense vegetation or forest, canopies

Forage plant: Not recorded

7. Great Zebra, *Graphium xenocles* Doubleday

Subfamily: Papilioninae

Genus: *Graphium*

Species: *xenocles*

Distinguishing Characters: Body black. Tailless. Upper dark brown with white stripes and spots. In *xenocles*, bright yellow sopt on tornus. Fore cell with white streaks , obliquely arranged, distal two are cojoined. Under dusty brown, paler towards apex, marks less prominent. Upperhind submarginal spots crescent shaped only 1-2. Wingspan: 90.1 mm. Not rare .

Global Distribution: Assam, Arunachal Pradesh, Nagaland, Myanmar.

Recorded Localities: Dehing Patkai forest track; Spring (April)

Land use land cover pattern: Open sunlight areas in forest

Forage plant: Not recorded

Tribe2: Papilioninae (Black bodied swallow tails)

8. Common Mime, *Papilio clytia clytia* Linnaeus (Form I: *clytia*)

Old name, *Chilasa clytia* Linnaeus

Subfamily: Papilioninae

Genus: *Papilio*

Species: *clytia*

Subspecies: *clytia*

Distinguishing Characters: Tailless. Antenna equal or less than half

costa. In *clytia clytia,* wings dark brown with white marks. Both upper and under hind tornus with black centered yellow spot. Upper hind with submarginal pale whitish marks diffused inwardly forming arrowhead like structure. Upperfore with rows of marginal and terminal marks. They serve as example of Batesian mimicry and appears like common crow. Wing span: 110mm

Global Distribution: India, Combodia, South Asia

Recorded Localities and season: Kokrajhar; Monsoon and post monsoon

Land use land cover pattern: Forest and open areas near forest, parks

Forage plant: Ixora , Hydrangea, Ixora

9. Common Mime, *Papilio clytia clytia* Linnaeus (Form II: *dissimilis*)

Old name: *Chilasa clytia* Linnaeus

Subfamily: Papilioninae

Genus: *Papilio*

Species: *clytia*

Subspecies: *clytia*

Distinguishing Characters: Body white and black, both upper and under hind tornus with black centered yellow spot. In *dissimilis,* upper fore with prominent pale discal spots and few apical streaks over black wings. Upperfore cell divided by three pale streaks and three marks end cell. Upper hind with long pale streaks. Upper hind with yellow marginal semilunar spots. Wings kept wide open while nectaring. Wing span: 90-120mm

Global Distribution: India, Combodia, South Asia

Recorded Localities and season: Kokrajhar ; Monsoon and post monsoon

Land use land cover pattern: Forest and open areas near forest, parks

Forage plant: Ixora spp., flowers of Horticultural plants

10. Common Mormon, *Papilio polytes romulus* Cramer (M)

Subfamily: Papilioninae

Genus: *Papilio*

Species: *polytes*

Subspecies: *romulus*

Distinguishing Characters: Black bodied swallow tail. Tailed. In *Papilio,* forewing vein2 straight, v7 is midway between v6 and v8. Male velvety black . In *polytes,* hind with a discal band of white large irregular shaped marks. Upper fore margin with a series of smaller white spots . Under hind possess white marginal and sub-marginal crescents . I have observed it flying very fast with fast wing beating around a female stichius form before mating. Wingspan: 90-100mm

Global Distribution: India

Recorded Localities and season: Dibrugarh , Tezpur, Guwahati, Kokrajhar; Monsoon and post monsoon

Land use land cover pattern: Open places near dense vegetation or forest , garden and parks.

Forage plant: *Ixora, Lanatana* spp., *Jasminum* spp., *Mussaenda* spp., Bougainvillea , Hydrangea

11. Common Mormon, *Papilio polytes* Linnaeus (F) (Form: *Cyrus*)

Subfamily: Papilioninae

Genus: *Papilio*

Species: *polytes cyrus*

Distinguishing Characters: Tailed . Upper brown with black streaks between veins. Elongated streaks in upper cell. In *cyrus,* upper fore with marginal white spots. Upperhind with broad discal white band. Hind with red crescent at tornus. Found while basking on a polyanthaus tree (December)

Recorded Localities and season: Dibrugarh University campus,

Biodiversity Park, Dibrugarh, Tezpur, Zoological garden, Guwahati, Kokrajhar; Monsoon and post monsoon

12. Common Mormon, *Papileo polytes* Linnaeus (F) (Form: *stichius*)

Subfamily: Papilioninae

Genus: *Papilio*

Species: *polytes*

Distinguishing Characters: The form *stichius* mimics the Common Rose. Body black, tailed . The forewing is velvety brown with black streaks between veins and no white marginal spots. The hind wing with central white streaks, submarginal series of red crescent in termen. Hind with white marginal marks between veins. It mimics the common rose in which the body is red. Wing span:90-100mm

Global Distribution: India

Recorded Localities and season: Dibrugarh , Tezpur, Guwahati, okrajhar; Spring,Monsoon

Land use land cover pattern: open places near dense vegetation or forest, garden and parks.

Forage plant: *Ixora, Lanatana* spp., *Jasminum* spp., *Mussaenda* spp., Bougainvillea , Hydrangea

13. Khasi Red Breasted Mormon, *Papilio alcmenor alcmenor* C and R Felder

Subfamily: Papilioninae

Genus: *Papilio*

Species: *alcmenor*

Subspecies: *alcmenor*

Distinguishing characters: Black bodied swallowtail. Upper bluish black. Tailless. Upper fore in male with red streak in cell at base. Cell

upperhind mostly without streak. Hindwing elongated. Forewing apical margin straight. Wing span:115-130mm

Global Distribution: India

Recorded Localities and season: Dibrugarh,Kaziranga, Ultapani reserve forest, Kokrajhar; Monsoon, post monsoon

Land use land cover pattern: Forest tracks, garden and parks.

Forage plant: *Ixora, Lanatana* spp., *Jasminum* spp., *Mussaenda* spp., *Bougainvillea, Hydrangea*

14. Great Mormon, *Papilio memnon agenor* Linnaeus

Subfamily: Papilioninae

Genus: *Papilio*

Species: *Memnon*

Subspecies: *agenor*

Distinguishing characters: Black bodied female swallowtail. In *Papilio*, forewing vein2 straight, v7 is midway between v6 and v8. In *memnon*, upper blue scaling against black background. Hind tornus with black centered red marks . They are highly polymorphic and several forms are found. In *agenor* , form -I, the tailed form uph has a discal row of white spot ,white marginal bars between veins, red lunule at tornus. In form-2, it is tailess male form with red basal triangular mark in underfore cell , red lunules underhind in both tornus and base. Wing span:120-150mm

Global Distribution: India, Burma, Nicobar, Myanmar, Bhutan

Recorded Localities and season: Jakai Biodiversity Park, Dibrugarh, Dhubri, Tezpur, Guwahati, Kokrajhar; Monsoon , post monsoon,winter

Land use land cover pattern: Forest tracks, open garden and parks.

Forage plant: *Ixora, Lantana* spp., *Jasminum* spp., *Mussaenda* spp., *Bougainvillea, Hydrangea, Clerodendrum*

15. Continental Great Mormon, *Papilio agenor agenor* Linnaeus

Subfamily: Papilioninae

Genus: *Papilio*

Species: *agenor*

SubSpecies: *agenor*

Distinguishing characters: Wings with blue streaks between black veins. Cells with streaks. Tailess male form may or may not have red basal triangular mark in fore cell , red lunules underhind in tornus and red marks at base. Upper hind tornus with black centered red lunule.

16. Great Mormon, *Papilio memnon alcanor* Linnaeus (F)

Subfamily: Papilioninae

Genus: *Papilio*

Species: *memnon alcanor*

Distinguishing characters: Black bodied female swallowtail. Upperfore, cell base red. In this *alcanor* form, upperfore greyish brown with black veins and streaks. Upperfore base of v1 and v2 have velvety black patch.

Upperhind cell with a white mark surrounded by elongated white marks outside cell. Tornus with black centered red spot. Red marginal marks between veins. Wing span:120-150mm

Global Distribution: India, Burma, Myanmar, Bhutan

Recorded Localities and season: Jakai Biodiversity Park, Dibrugarh, Dhubri, Tezpur, Guwahati, Kokrajhar; Monsoon , post monsoon,winter

Land use land cover pattern: Forest tracks, open garden and parks.

Forage plant: *Ixora, Lanatana* spp., *Jasminum* spp., *Mussaenda* spp., Bougainvillea, *Hydrangea, Clerodendrum*

17. Lime Butterfly, *Papilio demoleus* Linnaeus

Synonym: *Princep demoleus* Linnaeus

Subfamily: Papilioninae

Genus: *Papilio*

Species: *demoleus*

Distinguishing Characters: Body black with lemon yellow stripes. Wing without Tail. Marginal and submarginal rows of irregular lemon spots. In *demoleus,* FW with broad light yellow postdiscal bands broken into irregular patches. Hind tornus with red lunule bordered by black. Underhind with postdiscal brown marks, brownish lemon base with a lunule mid costa. Wingspan: 91mm

Global Distribution: India

Recorded Localities and season: Common across Assam

Land use land cover pattern: Found mostly on upper canopies in forests, bush levels in parks

Forage plant: *Ixora* spp., *Lantana camara* , *Citrus* spp., garden plants, *Hydrangea*

18. Himalayan Spangle, *Papilio protenor euprotenor* Fruhstofer (Form I)

Subfamily: Papilioninae

Genus: *Papilio*

Species: *protenor*

Subspecies: *euprotenor*

Distinguishing Characters: Black body, tailless. Head black with white spots. In *protenor,* bluish dusting on upperhind, upperfore with white

scales between blue dusted veins . In *protenor euprotenor,* under black veins with white scales between veins. Under hind with black centered red large lunule near tornus and apex. The basal red marks do not extend to base. Upperhind incomple red lunule at tornus. Observed during mud puddling. Wingspan:115mm

Global distribution: India, Burma.

Recorded Localities and season: Woody forest of Kaziranga, Dibrugarh; monsoon, Post monsoon

Land use land cover pattern: Forests and open space near forest

Forage plant: Not recorded

19. Himalayan Spangle, *Papilio protenor euprotenor* Fruhstofer (Form II)

Subfamily: Papilioninae

Genus: *Papilio*

Species: *protenor*

Subspecies: *euprotenor*

Distinguishing Characters: Black body, tailless. In *protenor,* blue dusting on upperhind, darker dusting upperfore. Upperhind lunule at tornus. In *protenor euprotenor,* under hind with black centered red large lunule near tornus and apex. Frequent wing beating during mud puddling. Wingspan:115mm

Global distribution: India, Burma

Recorded Localities and season: Fringe area of Dehing Patkai National Park, Dibrugarh ; monsoon

Land use land cover pattern: Forests and fringe areas

Forage plant: Not recorded

20. Patkai Common Raven, *Papilio castor castor* Westwood

Subfamily: Papilioninae

Genus: *Papilio*

Species: *castor*

Subspecies: *castor*

Distinguishing Characters: Body black with marginal white spots arranged in rows. Wing without tail. In *Papilio*, discocellular vein 6-7 almost straight. Hind with submarginal single row of spots extending upto halfway the anterior margin. Under fore submarginal two rows of

white spots . In *castor,* brown scales between veins. In *castor castor,* fore with single white spot at the distal end cell. Hind end cell with prominent white marks at space 4-6 . Wingspan:100-130mm

Global Distribution: India, Combodia, South Asia

Recorded Localities and season: Ultapani Reserve Forest, Kokrajhar, forest clearing; Summer

Land use land cover pattern: Forest tracks, recorded during mud puddling.

Forage plant: Not recorded

21. Common Raven, *Papilio castor polias* Jordan (M)

Subfamily: Papilioninae

Genus: *Papilio*

Species: *castor*

Subspecies: *polias*

Distinguishing features: Black bodied swallowtail. In *Papilio,* discocellular vein 6-7 almost straight. Wings without tail. In *castor,* brown scales between black veins. In, *castor polias,* upper hind wing with large white mark starting vein4. No submarginal spots. Cell with prominent longitudinal black veins. Omnivorous and can act as scavengers. Wing span:

Global Distribution: India, Sikkim, Nepal, Bhutan

Recorded Localities and season: Panbari , Kaziranga;

Land use land cover pattern: Woody forest and moist habitats

Forage plant: Not recorded

22. Khasi Yellow Helen, *Papilio nephelus chaon* Westwood

Subfamily: Papilioninae

Genus: *Papilio*

Species: *nephelus (chaon)*

Subspecies: *chaon*

Distinguishing Characters: Body black with white spots. Hind wings with tail. In Helenus, upper fore with no prominent discal band. Upper hind with white patch of marks. In *nephelus chaon*, under hind with four white discal patches starting from vein 4. Small yellow discal spots in space1-3. Under hind with submarginal yellow and white lunules. Wing span: 115-130mm

Global Distribution: India, Combodia, South Asia

Recorded Localities and season: Ultapani Reserve Forest, Kokrajhar, Udalguri; Summer

Land use land cover pattern: Forest tracks. Recorded during mud puddling , basking and nectaring

Forage plant: Lantana Camara, Ixora, Hydrangea

23. Red Helen, *Papilio helenus* Linnaeus

Subfamily: Papilioninae

Genus: *Papilio*

Species: *helenus*

Distinguishing characters: Body black. Upper hind with three white marks in v6,7,8 . It is similar with yellow helen except that the white patch upperhind is divided by two veins instead of three and the submarginal spots underhind are dull red in colour. Wing span: 115-130mm

Global Distribution: India, Combodia, South Asia

Recorded Localities and season: Jakai Biodiversity Park, Dibrugarh, July

Land use land cover pattern: Forest tracks. Recorded during basking and nectaring

Forage plant: Lantana Camara, Ixora, Hydrangea

24. Common Peacock, *Papilio polyctor polyctor* Boisduval

Subfamily: Papilioninae

Genus: *Papilio*

Species: *polyctor*

Sub species: *polyctor*

Distinguishing characters: Body black. Wing with tail. Upper black with metallic green scales. In *polyctor polyctor*, in upper fore, no streak in cells. Upper fore with very narrow discal metallic green band. Upper hind with prominent iridescent blue patch. Under hind rows of semilunar submarginal reddish lunules. Now the east Himalayan species is proposed as *bianor* [1,5] Wing expanse: 120-150mm

Global Distribution: India, Nepal, Bhutan,Sikkim, Myanmar, and Pakistan.

Recorded Localities and season: Kaziranga

Land use land cover pattern: woody and open tracks near forests

Forage plant: Not recorded

Tribe3: Troidinae (Birdwing, Batwing, Roses, Windmills)

25. Khasi common Birdwing, *Troides helena cerberus* C&R Felder

Subfamily: Papilioninae

Genus: *Troides*

Species: *helena*

Subspecies: *cerberus*

Distinguishing Characters: Red bodied swallowtails due to presence of red marks underside thorax. In *Troides*, heavy yellow body with black marks. Neck has a red ring. Upper cell black. HW bright shining yellow with black veins, termen with black margin produced into triangles in spaces between veins. In *helena*, upper fore no pale mark end cell. In *cerberus*, upper hind base 1b yellow. Wing span: 148 mm.

Global Distribution: India

Recorded Localities and season: Jakai Biodiversity Park, Dibrugarh, Panbari Reserve Forest, Kaziranga, Dehing Patkai Reserve forest ; Monsoon

Land use land cover pattern: Forest patches

Forage plant: *Ixora* spp., *Lantana camara, Hydrangea*

26. Khasi Golden Birdwing *Troides aeacus aeacus* C&R Felder

Subfamily: Papilioninae

Genus: *Troides*

Species: *aeacus*

Subspecies: *aeacus*

Distinguishing characters: Prominent whitish bordered veins both upper and underfore. In *Troides*, heavy yellow body with black marks. Upper, neck with a red ring. Under red marks on either side of thorax. In *aeacas aeacus*, the fore cell with almost half with white striped veins. In female, series of submarginal wedge shaped black spots. Base of 1b black. Protected under Schedule II of IWPA,1972. Wing span:150-170mm

Global Distribution: India

Recorded Localities and season: Dehing Patkai Reserve forest ; Post-monsoon (August)

Land use land cover pattern: Forest clearings, forests

Forage plant: *Ixora* spp., *Lantana camara, Hydrangea*

27. Common Batwing, *Atrophaneura varuna astorion* Westwood

Subfamily: Papilioninae

Genus: *Atrophaneura*

Species: *varuna*

Subspecies: *astorion*

Distinguishing characters: Red bodied swallowtail. Upper hind bluish black unmarked and dorsal scent fold is rounded. Upperfore with pale streaks between veins. Wing with tail. Wing expanse: 88-136 mm

Global Distribution: India, Nepal, Bhutan.

Recorded Localities and season: Near Dehing Patkai National Park

Land use land cover pattern: Tracks near forests

Forage plant: Not recorded

28. East Himalayan Great Windmill, *Byasa dasarada dasarada* Moore

Subfamily: Papilioninae

Genus: *Byasa*

Species: *dasarada*

Subspecies: *dasarada*

Distinguishing characters: Red bodied swallowtail. Upper hind red submarginal spots irregular. Spot in 4 is white, spot in 2,3 are with white margin. Underhind, large spot in 5 and also a spot in 6. Wing expanse: 100-140mm

Global Distribution: India, Nepal, Bhutan,Sikkim, Myanmar, and

Pakistan.

Recorded Localities and season: Fringe area of Dehing Patkai National Park

Land use land cover pattern: Near forests

Forage plant: Not recorded

Family: Nymphalidae

Dipsikha Bora

The butterflies under the family Nymphalidae are called browns because of their ground colouration and also brush-footed because of their incompletely developed forelegs. Those non-functional legs remain folded on the thorax and the short tibiae are clothed with hairs. The Nymphalid collections presented in this chapter includes subfamilies Danainae, Satyrinae, Heliconinae, Limenitinae, Charaxinae Nymphalinae, Cyrestinae and Biblidinae. In Evans (1932), the browns and rings are classified under another family Satyridae, but in the updated classification (Imms,1977), Satyridae is included as sub-family, Satyrinae under Nymphalidae. Satyrinae includes the browns which can be recognized by swollen veins at the base of the forewings or eye like rings. They are small to medium sized butterflies, shade lovers, brown or tawny in colour and may exhibit cryptic colouration on the undersurface. Satyrinae includes the rings, evening browns and palmflies. The rings are named commonly based on the number and positions of ocelli present on the underside of the hind wings. These ocelli are analyzed to show convergent evolution from the wing rings of Mid-Mesozoic kalligrammatid lacewings (Neuroptera). Danainae was also considered by Evans (1932)as a separate family, Danaidae. In Danaids, antenna is filiform or clubbed, claws bear appendages, hind vein 8 is not parallel to vein 7 and meets base about the origin of vein 7. The subfamily includes the tigers and crows. The caterpillars of these butterflies exhibit the phenomenon of acquired defense by sequestering and transferring cardiac glycosides from host plants to the adults for defense against predators and parasites. The subfamily Charaxinae includes Rajah and Nawabs equipped with a large thorax and narrow pointed wings. Rajahs and Nawabs may be sighted commonly puddling in mud, small rain water pool, carrion in forest clearings. The subfamily nymphalinae includes different groups of Nymphalids including Sailers and Sergeants, Egg flies, Fritillary, Pansys and Castors. Heliconiinae includes Lacewings. Acraeinae includes the costers. These butterflies belonging to Nymphalinae and Heliconinae are commonly sighted in gardens having flowers for nectaring, arenas for basking and shades for rest.

The family is the largest of all butterfly families with about 6000 described species across the globe out of which over 520 species are recorded in India. In this book we are reporting 67 commonly observed nymphalids including various forms. Mimicry is observed between Dannaid great egg fly and common tiger . Some are known for their migratory behaviour .

NYMPHALIDAE

Subfamily 1: Danainae (Tigers, Crows)

1. Oriental Plain Tiger, *Danaus chrysippus chrysippus* Linnaeus

Subfamily: Danainae

Genus: *Danaus*

Species: *chrysippus*

Subspecies: *chrysippus*

Distinguishing Characters: Tawny or orange body and wing with black and white scales. In *Danaus,* antenna clubbed. In *chrysippus,* veins not black. Upper fore costal margin, apex upto tornus black. The black apex is with band of white patches. Upper fore and hind black margins are with white spots. Underside dull. End cell in hind has black mark. Wing span: 75mm

Global Distribution: India, Myanmar, Bangladesh, Andaman, Tropical Africa, Southern Europe, Indo-australia, Oriental region

Recorded Localities and season: Dibrugarh; Pre and Post monsoon

Land use and land cover pattern: Open gardens and parks, Forest tracks.

Forage plant: *Celosia, Lantana,* flowers under Asteraceae

2. Oriental Striped Tiger, *Danaus genutia genutia* Cramer

Subfamily: Danainae

Genus: *Danaus*

Species: *genuitia*

Subspecies: *genuitia*

Distinguishing Characters: Tawny or orange wings with black stripes. Antenna clubbed. In *genutia,*veins are black. Under and upper hind paler. The wings on both sides have black borders with two rows of spots. The forewing apex is black with band of white patches. Males have a white centered black patch on hindwing outside cell which contains scent

glands. It is called as androconia. Wingspan: Male:93mm; Female 89mm

Global Distribution: India, South East Asia

Recorded Localities and season: Dibrugarh (University Botanical Garden); Pre monsoon (March)

Land use and land cover pattern: Open gardens and parks

Forage plant: *Celosia, Lantana*

3. Dark Glassy Tiger, *Parantica agleoides* C&R Felder

Sub family: Danainae

Genus: *Parantica*

Species: *agleoides*

Distinguishing Character: Dark. In *Parantica* antenna gradually clubbed. In *agleoides*, cells in both wings pale with two black streaks. One black streak in 1b. Both wings with many white marks. Pale streak from base cell along costa. UP and UN exhibit same pattern except scent scales like dark black patch UNH near tornus in male. Pale areas extensive. Upper both wings with rows of marginal and submarginal white spots. Wingspan: 89mm

Global Distribution: India, Sri Lanka, Bhutan

Recorded Localities and season: Namrup, near Dehing Patkai National

Park

Land use and land cover pattern: Gardens, Forest edge ; Forage plant: Not recorded

4. Himalayan Glassy Tiger, *Parantica aglea* Stoll

Sub family: Danainae

Genus: *Parantica*

Species: *aglea*

Distinguishing Characters: Black with pale blue to white transparent markings. In *aglea,* upper fore cell with a bluish white streak with a black apical line. The white streak is unbroken transversely. The streak below cell is much broader. Pale streak from base cell along costa. Upper fore outside cell towards apex has two pale streaks. Besides, many more pale marks are giving a pattern to the wings. Antenna clubbed. Wingspan: 100mm

Global Distribution: India, Burma, Bangladesh, Bhutan

Recorded Localities and season: Dibrugarh University campus; Monsoon (August)

Land use and land cover pattern: Open gardens , Parks, Forest tracks

Forage plant: *Ageatum conizoides*

5. Oriental Blue Tiger, *Tirumala limniace exoticus* Gmélin (M)

Sub family: Danainae

Genus: *Tirumala*

Species: *limniace*

Subspecies: *exoticus*

Distinguishing Characters: In *Tirumala,* antenna narrowly clubbed . Wings brown with bluish narrow streaks and spots. Fore wing upper basal streak in cell is short. In *limniace,* the mark end cell below streak is large. Fore, no pale streak from cell base along costal vein. UNH with forked blue streak in cell. Under wings with marginal lines and submarginal spots between veins. Scent scale pouch present . Wingspan: 90-100mm

Global Distribution: India, Nepal, Bhutan, Sri Lanka

Recorded localities and season: Kaziranga; Dehing Patkai range; Pre monsoon

Land use and land cover pattern: Woodland, forest edge, Open gardens, parks

Forage plant: *Crotolaria* spp., *Zinnia, Lantana*

6. Dark Blue Tiger, *Tirumala septentrionis* Butler

Sub family: Danainae

Genus: *Tirumala*

Species: *septentrionis*

Distinguishing Characters: Large. In *Tirumala,* antenna narrowly clubbed. In *septentrionis,* upper with dark blue narrow markings of spots and bars. Underhind cell with a v-shaped mark of two blue bars. Under dull brown with lighter blue marks. Dark pouch underhind. Observed moving on the wings rarely sitting. Wingspan: 90-105mm

Global Distribution: India, Nepal, Bhutan, Sri Lanka

Recorded Localities and season: Jakai Biodiversity Park, Dibrugarh,

Spring (May)

Land use and land cover pattern: Forest edge, Open gardens ,

Forage plant: Asteraceae

7. Blue Glassy Tiger, *Ideopsis similis* Linnaeus

Sub family: Danainae

Genus: *Ideopsis*

Species: *similis*

Distinguishing Characters: Upper dark brown with blue markings. Under dull brown with blue markings. In fore, the blue mark in cell is broad and divided by a black bar. In hind the blue mark in cell is v- shaped. Wingspan: 75-85mm

Global Distribution: India, Myanmar, Sri Lanka

Recorded Localities and season: Jakai Biodiversity Park, Dibrugarh, Dehing Patkai range ;Spring (May)

Land use and land cover pattern: Forest edge, Open gardens ,

Forage plant: *Lantana camera*

8. Blue King Crow, *Euploea klugii klugii* Moore

Sub family: Danainae

Genus: *Eoploea*

Species: *klugii*

Subspecies: *klugii*

Distinguishing characters: Large. Upper brown with blue colour on cells and discal areas. In *Euploea*, vein 8 and vein 9 almost parallel and are away from costa. Antenna clubbed. In *klugii klugii*, upper cell contains one blue spot and elongated blue spots outside cells. Upper both with marginal rows of spots and submarginal rows of elongated marks. Under dark brown with two rows of marginal and submarginal white spots. No spot in underhind cells. Wing span: 110mm

Global Distribution: India, Myanmar.

Recorded Localities and season: Kokrajhar, Kaziranga; monsoon

Land use and land cover pattern: Woodland, forest tracks, gardens

Forage plant: *Dahlia,* Marigold (*Tagetes*) *etc.*

9. Magpie Crow, *Euploea radamanthus* Fabricius

Sub family: Danainae

Genus: *Euploea*

Species: *radamanthus*

Distinguishing characters: Black with cream coloured or pale patches. Fore wing vein 8 and vein 9 almost parallel. In *radamanthus,* a large mark end cell in both upper and under. Both upper and under hind with many pale streaks and except two streaks on posterior hind wing, other streaks are not extended to base. UPF with blue streak outside cell and two rows of blue marginal spots UPF. Wing span: 91 mm

Global Distribution: India, Myanmar.

Recorded Localities and season: Dehing Patkai, Autumn

Land use and land cover pattern: Woodland, forest track, gardens

Forage plant: *Dahlia,* Marigold (*Tagetes*)

10. Common Crow, *Euploea core* Cramer

Sub family: Danainae

Genus: *Euploea*

Species: *core*

Distinguishing characters: Upper dark velvety brown, under pale. Under fore with white spot end cell. Under with two rows of marginal cells. Upper fore submarginal and discal spots prominent. Wing span: 90 mm

Global Distribution: India, Myanmar.

Recorded Localities and season: Dibrugarh University Campus, Autumn, Sivsagar

Land use and land cover pattern: Woodland, forest tracks, gardens

Forage plant: *Dahlia,* Marigold (*Tagetes*)

11. Striped Blue Crow, *Euploea mulciber* Cramer

Sub family: Danainae

Genus: *Euploea*

Species: *mulciber*

Female

Distinguishing Characters: Upper dark brown with blue glossy area. In *Euploea,* vein 8 and vein 9 almost parallel and are away from costa. Antenna clubbed. In female, upper glossy area is small. Upperfore with marginal, postdiscal and discal white spots. Upperfore cell with a white indistinct streak followed by a white spot. Upper hind with white streaks radiating from base to submarginal area. Marginal white spots between veins upper hind. Under is paler brown with similar pattern of white spots and stripes. In *mulciber mulciber,* female underhind entirely white streaked. Protected under Schedule IV of IWPA, 1972. Wing span: 90-100 mm

Global Distribution: India, Burma, Andaman

Recorded Localities and season: Dehing Patkai; July

Land use land cover pattern: Forest clearings

Forage plant: *Ageratum conyzoides*

Sub family2: Satyrinae (Browns, Rings, Palm Flies)

12. Common Evening Brown, *Melanitis leda* Linnaeus (WSF)

Sub family: Satyrinae

Genus: *Melanitis*

Species: *leda*

Distinguishing Characters: Large. Body and wings brown. Upper fore with dark brown spot bounded by orange boundary and the eyespot has

one central and one upper marginal white spot. Fore wing termen with angle. Hind wing is with a tooth and termen straight to tornus. Under dull brown with black edged small spots indicating its wet season form of the specimen. In *Melanitis,* white spot is present on upper outer edge of dark brown ocellus and the veins at the base are not swollen. In the botanical garden of Dibrugarh University we collected two forms in premonsoon period . Underside appear like dry leaves. Wing span: 90mm

Global Distribution: India, Burma, Andaman

Recorded Localities and season: Assam; Post Monsoon

Land use land cover pattern: Mostly on grasses in open land or gardens

Forage plant: Not recorded

13. Common Evening Brown, *Melanitis leda* Linnaeus (DSF)

Distinguishing characters: Forewing produced and prominently angled at v5(Female). Upper brown with lunules. Hindwing tailed at v1 &3. Pupils underhind in the DSF almost obscure. Veins pale, under bear no brown streaks.

Recorded Localities and season: Dibrugarh University, Dibrugarh, February

Land use land cover pattern: Mostly on grasses and dry litters in open land or gardens

14. Oriental Common Evening Brown, *Melanitis leda leda* Linneaus

Sub family: Satyrinae

Genus: *Melanitis*

Species: *leda*

Subspecies: *leda*

Distinguishing Characters: Body and wings brown. Hind wings toothed. In *Melanitis,* upper fore with white double pupiled dark brown spots towards apex and with a white spot on upper outer edge. Veins at the base not swollen. UPH with three marginal dark brown pupil with white spot. Under dull brown with dark brown striations. UNH with six rings with the costal being the largest. UNF with three or four small rings. Wing span: 60mm

Global Distribution: India, Burma, Andaman

Recorded Localities and season: Assam; Pre Monsoon

Land use land cover pattern: Grasses, open gardens

Forage plant: Not recorded

15. Himalayan Common Bush Brown, *Mycalesis perseus blasius* Fabricius (WSF)

Sub family: Satyrinae

Genus: *Mycalesis*

Species: *perseus*

Subspecies: *blasius*

Distinguishing Characters: Brown in colour. Under fore and hind with a whitish brown brands which lie in line when wings are closed. White centered, brown bordered black eye spots, two between the white brand and apical margin of UNF . In *Mycalesis*, under, the discocellular vein is angled. Forevein 12 is swollen. In *mineous* group, the dcv underhind is prominently angled at v3. In *perseus,* underhind bears series of seven eye spots of different size with the one in 3,4,5 shifted out of line. Upper fore has single eye spot. Two forms are observed. In form I, underhind, the pale border of the small apical eyespots are joined while in form II, they are separate. Wing span: 38-50 mm

Global Distribution: India, Myanmar, Sri Lanka.

Recorded Localities and season: All throughout Assam; Monsoon

Land use and land cover pattern: Mostly on herbs and shrubs

Forage plant: Not recorded, found while mud puddling and also attracted to fruit bait.

16. Himalayan Common Bush Brown, *Mycalesis perseus blasius* Fabricius (DSF)

Sub family: Satyrinae

Genus: *Mycalesis*

Species: *perseus*

Subspecies: *blasius*

Distinguishing Characters: Body and wings dark brown. Forewing with costa rounded. Upper fore with dark brown pupilled ocellus in a pale region, without any yellow ring. UPH with dark brown fine marginal lines, no prominent ringed ocellus. In *Mycalesis,* the angle of cell towards apex is acute and vein five on UPF is swollen at base. Upper hind and under fore with brand and tuft of hair. In *perseus blasius,* upperfore pupil at vs2. UNF in DSF dark, bears very small ocelli 2-5 forming a curve and looks like shady dry leaf . Wing span: 45-55mm

Global Distribution: India, Myanmar

Recorded Localities and season: Dibrugarh, riverside park, found them in garden near dump of dry leaves and garbage; Winter (December)

Land use land cover pattern: Leaf litters

Forage plant: Not recorded

17. Dark-Branded Bush Brown, *Mycalesis mineus polydecta* Cramer

Sub family: Satyrinae

Genus: *Mycalesis*

Species: *mineus*

Sub species: *polydecta*

Distinguishing Characters: Dark Brown in colour. Under fore and hind with a whitish brands . White centered , brown bordered black eye spots, two UNF and seven UNH . In *Mycalesis*, under, the discocellular vein is angled. Fore vein 12 is swollen. In *mineus*, underhind in 1a,1b,2,3 are in line . In *mineus polydecta*, UPF with prominent pale area bearing eyespot, the outer yellow ring of which diffuses into the pale area. A golden brown brand with prominent androconia upperhind and a smaller brown brand underfore. Wing span: 42mm

Global Distribution: India, Myanmar, Sri Lanka.

Recorded Localities and season: All throughout Assam; Monsoon

Land use and land cover pattern: Mostly on herbs and shrubs

Forage plant: Not recorded, found attracted to fruit bait

18. Dark Branded Bush Brown, *Mycalesis mineus mineus* Linnaeus

Sub family: Satyrinae

Genus: *Mycalesis*

Species: *mineus*

Subspecies: *mineus*

Distinguishing Characters: Brown in colour.. White centered , pale brown bordered black eye spots, two UNF and five UNH . In *Mycalesis,* under, the discocellular vein is angled. Fore vein 12 is swollen at base. Veins under are dark brown in colour. In *mineus mineus,* underhind spots in 1a,1b,2,3 are in line. Under fore and upper hind with brown brands, upper hind bearing a tuft of hair. UPF ocellus in vs 2, underfore a prominent ocellus in vs 2 and a smaller one in vs5. Underhind, the eyespot in vs 2 is the largest. UPF the pale boundary of ocelli is narrow. Occur during early morning or crepuscular period. Wing span: 44mm.

Global distribution: India, Burma

Recorded Localities and season: Dibrugarh; Post Monsoon

Land use and land cover pattern: Mostly on understory grasses, open gardens

Forage plant: Not recorded, found attracted to fruit bait

19. Common Four Ring, *Ypthima huebneri* Kirby

Sub family: Satyrinae

Genus: *Ypthima*

Species: *huebneri*

Distinguishing Characters: With rings. In *Ypthima,* lower discal vein straight. Brown in colour. In *Ypthima huebneri,* UPH with ocelli in 2,3. UNH with three yellow ringed eyespots in line along termen in vein1, 2, 3. UNH with one additional apical eye spot. Under fore with one larger sized eyespot. Upper fore with one apical eyespot, upperhind with two large eyespots and one small tornal eyespot. Upperhind with two marginal eyespots. Upper darker brown marginal border narrow. Wing span: 30-40mm

Global Distribution: India, Myanmar.

Recorded Localities and season: All throughout Assam; Monsoon, Post monsoon

Land use and land cover pattern: Mostly on grasses and shrubs

Forage plant: Flowers of herbs like those from compositae family.

20. Common Five Ring, *Ypthima baldus baldus* Fabricius

Sub family: Satyrinae

Genus: *Ypthima*

Species: *baldus*

Species: *baldus*

Distinguishing Characters: Upper brown with rings. In *Ypthima,* lower discal vein straight. Upper fore with a prominent double pupiled eyespot with yellow border. Under whitish with coarse dark brown striations. In *baldus baldas,* UNH three pairs of ocelli along the termen are not in line . In *baldus baldus,* Underhind, three pairs of ocelli, middle pair larger and out of line. Upperhind has two prominent eyespots and thick layer of hair in basal part.

Global Distribution: India

Recorded Localities and season: Dibrugarh; Spring (May)

Land use and land cover pattern: Urban gardens, parks

Forage plant: Seasonal flowers.

21. East Himalayan Five Ring, *Ypthima sakra sakra* Moore (Form I)

Sub family: Satyrinae

Genus: *Ypthima*

Species: *sakra*

Subspecies: *sakra*

Distinguishing Characters: With rings.Upper brown. In *Ypthima,* lower discal vein straight. Upper fore with dark marginal band and a prominent double pupiled eyespot with pale border . Under whitish with coarse dark

brown striations. In *sakra,* the upperhind four eyespots are said to be present and sometimes with only two. This form is with two prominent eyespots and one small tornal eyespot. Underhind with three pairs of ocelli, the yellow rings coalesced in each pair.. Attracted to fruit bait also. Frequently folds and open wing at the time of nectaring. Wing span: 30-40mm

Global Distribution: India, Myanmar.

Recorded Localities and season: All throughout Assam; Monsoon, Post monsoon

Land use and land cover pattern: Mostly on grasses and shrubs

Forage plant: Flowers of herbs like those from compositae family, Ixora, Catharanthus, seasonal flowers.

22. East Himalayan Five-Ring, *Ypthima sakra sakra* Moore (Form II)

Sub family: Satyrinae

Genus: *Ypthima*

Species: *sakra*

Sub species: *sakra*

Distinguishing Characters: Upper brown with rings. In *Ypthima,* lower discal vein straight. Upper fore with a prominent double pupiled eyespot

with yellow border . Under whitish with coarse dark brown striations. In *sakra*, the upperhind four eyespots are said to be present and sometimes with only two. In this form upperhind is with one small apical spot and two large adjascent spots and thick layer of hair in basal part. Underhind with three pairs of ocelli, the yellow rings coalesced in each pair. Found attracted to fruit bait . Frequently folds and open wing at the time of nectaring. Wing span: 45-50mm

Global Distribution: India, Myanmar.

Recorded Localities and season: Dibrugarh; Monsoon, Post monsoon

Land use and land cover pattern: Urban gardens, parks

Forage plant: *Catharanthus,* seasonal flowers.

23. Common Five-Ring, *Ypthima baldus* Fabricius (DSF)

Sub family: Satyrinae

Genus: *Ypthima*

Species: *baldus*

Distinguishing Characters: Upper brown with rings. In *Ypthima*, lower discal vein straight. Under whitish with coarse dark brown striations. In *baldus*, UNH five ocelli along the termen are not in line and present like specs only. UNF with double pupiled ocellus.

Global Distribution: India

Recorded Localities and season: Dibrugarh; Winter (February)

Land use and land cover pattern: Urban gardens, parks

Forage plant: Seasonal flowers.

24. Himalayan Common Palmfly, *Elymnias hypermnestra undularis* Drury

Sub family: Satyrinae

Genus: *Elymnias*

Species: *hypermnestra*

Subspecies: *undularis*

Male

Female

Distinguishing Characters: In *Elymnias,* hind with prediscoidal cell at base of v8. Antenna gradually clubbed. In *hypermnestra,* UPH with ochreous marginal brand. In *undularis,* UPF with coffee brown with blue gloss and no brand. Under fore with apical pale triangular mark. Under wings with white striations. Hind wings toothed. Rests with wings folded. Wingspan 72mm

Global Distribution: India, Myanmar, Sri Lanka, Andaman, Pakistan

Recorded Localities and season: Assam (Guwahati, Dibrugarh University campus, Kaziranga); Observed during spring, Post monsoon

Land use and land cover pattern: Mostly on shrubs

Forage plant: Garden palms

25. Peal's Palmfly, *Elymnias pealii* Wood-Mason

Sub family: Satyrinae

Genus: *Elymnias*

Species: *pealii*

Distinguishing Characters: In *Elymnias,* hind with prediscoidal cell at base of v8. Body and UND wing Chestnut red. Under wings with many white striations. In *pealii* , UNH in space 7 is prominent white spot . Hind wings toothed. Protected under schedule I of IWPA 1972 Wingspan: 75-85

Gobal Distribution: India, Myanmar, Sri Lanka, Andaman, Pakistan

Recorded Localities and season: Assam(Guwahati, Dibrugarh University campus, Kaziranga); Monsoon, Post monsoon

Land use and land cover pattern: Mostly on shrubs

Forage plant: Garden palms

26. Spotted Palmfly, *Elymnias malelas* Hewitson

Sub family: Satyrinae

Genus: *Elymnias*

Species: *malelas*

Distinguishing Characters: In *Elymnias,* hind with prediscoidal cell at base of v8. Fore above with blue sheen and spots. In *malelas* , underfore with white spot in cell. Underhind with white spots. Wing margin with white marks. Under hind dull brown with dark brown veins and white

striations. Wingspan: 80-85mm

Global Distribution: India, Myanmar, Sri Lanka, Andaman, Pakistan

Recorded Localities and season: Sivasagar; August

Land use and land cover pattern: Mostly in banana cultivation

Forage plant: Not recorded

27. Nigger, *Orsotriaena medus medus* Fabricius

Sub family: Satyrinae

Genus: *Orsotriaena*

Species: *medus*

Subspecies: *medus*

Distinguishing Characters: Wing colour is van dyke brown. Under white discal line running through both the wings and fall in the same line when wings are folded. Under fore with two white centered eye spots of equal size. Under hind with similar two eye spots of equal size and a smaller one close to costa. Both wings are bordered with darker margin. In *Orsotriaena,* in fore the vein10 from cell . In *medus,* underhind sub, marginal line prominent, conjoined ocelli at apex. Weak flight. Wing span: 45-52mm

Global Distribution: India, Myanmar, Sri Lanka, Andaman.

Recorded Localities and season: Assam; Post monsoon

Land use and land cover pattern: Mostly on herbs and shrubs

Forage plant: Not recorded

Subfamily 3: Heliconiinae (Coster, Yeoman, Fritillary, Leopard)

28. Leopard Lacewing, *Cethosia cyane* Drury

Sub family: Heliconinae

Genus: *Cethosia*

Species: *cyane*

Distinguishing Characters: Body tawny. In *Cethosia,* cells closed. In *cyane,* upper wings with white band across black apical patch and extends beyond cell. Tawny areas of wings bear black spots. Black marginal border of wings with v- shaped white markings. Basks with wings spread. Wing span: 80-95mm

Global Distribution: India, Myanmar, Nepal, Bhutan

Recorded Localities and season: Dibrugarh; Post Monsoon (October)

Land use and land cover pattern: Open Gardens

Forage plant: *Catharanthus* flowers.

29. Indian Fritillary, *Argyreus hyperbius hyperbius* Linnaeus

Sub family: Heliconinae

Genus: *Argyreus*

Species: *hyperbius*

Sub species: *hyperbius*

UP (Female)

Distinguishing Characters: Upper tawny with black spots and marks and under usually with silvery marks. In *hyperbius*, under ochreous brown. Apex with white band. Upperhind with submarginal row of white centered brown spots. Wing margins between veins curved with black and tawny border giving the appearance of designer lace. In *hyperbius hyperbius*, fore apex much produced. Fore, v2 swollen. Status is 'Not rare'. Wing span: 80mm

Global Distribution: India, Myanmar.

Recorded Localities and season: Dibrugarh (University); Premonsoon Monsoon

Land use and land cover pattern: Gardens

Forage plant: *Zinnia*

30. Common Leopard, *Phalanta phalantha* Drury

Sub family: Heliconinae

Genus: *Phalanta*

Species: *phalantha*

Distinguishing Characters: Upper tawny with dark brown spots and marks and under dull. In *Phalanta*, upper fore cell with five lines including end bar. Margin with wavy markings and submargin with row of rounded marks. Underhind light yellow with dark brown and reddish marks. Underhind bears orange crowned silver marks in 1b-3. Observed basking under sun in the month of January and active during May. Wing span: 50-60mm

Global Distribution: India, Myanmar, Sri Lanka

Recorded Localities and season: Jakai Biodiversity Park, Dibrugarh, Sivasagar ; Winter to Premonsoon

Land use and land cover pattern: Gardens

Forage plant: Seasonal flowers for nectaring, Asteraceae

31. Large Yeoman, *Cirrochora aoris* Doubleday

Sub family: Heliconinae

Genus: *Cirrochora*

Species: *aoris*

Distinguishing Characters: Upper bright tawny with black spots and marks and under dull with silvery or white marks. Cell end with a prominent black curvy line. Upper with marginal black curvy line, hind with black spots between veins. In *Cirrochora,* lower end of discocelular vein ends before start of v3 in forewing. In *aoris,* upperfore apex narrowly black, darkbar end cell double. Under, broad white discal band followed by basal narrow white band. The white discal band underfore is not of uniform widh. Upper fore apex slightly produced but truncate. Wing span: 80-90 mm

Global Distribution: India (Assam, Sikkim, Arunachal Pradesh,, Myanmar, Bangladesh, Srilanka)

Recorded Localities and season: Dehing-Patkai forest clearing ; May

Land use and land cover pattern: forest tracks

32. Yellow Coster, *Acraea issoria* Hubner

Sub family: Heliconinae

Genus: *Acraea*

Species: *issoria*

Distinguishing Characters: Upper yellow with black wavy marginal marks bordering yellow spots. Under dull yellow with black wavy marginal marks. Underhind with yellow submarginal spots between veins making a yellow line. Male with end bar in cell. Wing span: 50-70 mm

Global Distribution: India (Assam, Arunachal Pradesh), Myanmar, Bhutan

Recorded Localities and season: Jokai Biodiversity park, Dibrugarh,

Dehing-Patkai forest clearing ; Winter, Premonsoon (Feb)

Land use and land cover pattern: forest tracks

Nectaring: *Lantana camera*

Subfamily 4: Limenitidinae (Baron, Sergeants, Sailer)

33. Khasi Dark Archduke, *Lexius dirtea khasiana* Swinhoe (=intermedia Tyt.) (F)

Sub family: Limenitinae

Genus: *Lexius* (= *Euthalia*)

Species: *dirtea*

Subspecies: *khasiana (=intermedia)*

Distinguishing Characters: Upper fore base 2 with four spots followed by one elongated mark. Upper fore marks yellow, under wing brushed with blue, spots whitish. In *Lexius,* cells are found open. In *dirtea khasiana* (=*intermedia*) female, spots upper fore pale yellow. Spot at base of 2 is not too small. Marginal spots are dark yellow in both sides of the wing. Antenna tip black gradually clubbed. Included under Schedule II of IWPA 1972.Wingspan: 90 mm

Global Distribution: India, Nepal, Bhutan, Sri Lanka

Recorded Localities and season: Dibrugarh University Campus, Sibsagar, Jakai Biodiversity Park, Dibrugarh; Autumn

Land use and land cover pattern: Open gardens, parks, damp places

Forage plant: Fig fruits

34. Yellow Tipped Archduke, *Lexius pardalis* Moore

Sub family: Limenitinae

Genus: *Lexius* (= *Euthalia*)

Species: *pardalis*

Distinguishing Characters: Upper fore base 2 with four whitish spots followed by one elongated mark. Upper hind with bluish marks. In *Lexius,* cells are found open. In *pardalis,* the antennae of both sexes are yellowish tipped with orange hue, not black. In female, bluish white spots are the distinguishing character. Spot at base of 2 is not too small. Antenna gradually clubbed. In male, upper velvet dark brown . Upper fore with bluish specks along costa and a prominent white spot at apex. Upper hind broad blue marginal border bearing small black spot between veins. Upper fore narrow blue marginal border not reaching apex. They were found in damp places with scattered rotten fruits. Protected under Schedule II of IWPA 1972. Wingspan: 90mm

Global Distribution: India, Nepal, Bhutan, Sri Lanka

Recorded Localities and season: Jakai biodiversity Park , Dibrugarh, Sibsagar, Charaideo, Dehing Patkai, Summer,Autumn

Land use and land cover pattern: Open gardens , parks, damp places

Forage plants: Fruits of figs, bhumura, Sam Kaothal

35. Great Archduke, *Lexius cyanipardus* Butler

Sub family: Limenitinae

Genus: *Lexius*

Species: *cyanipardus*

Female

Male

Distinguishing Characters: In *Lexius*, cells are found open. In *cyanipardus*, antenna black in both sexes. Male upperfore velvety black with spots in cell and along costa. Upperhind blue margin is broad and bear black prominent spots. Upperfore blue marks are narrow and not extended to apex. In female, upper dark brown with pale spots and marks in fore and bluish in hind. Spot at base of 2 is very small. *Lexius c. cyanipardus* is protected under schedule II of IWPA1972 Wingspan: 110mm

Global Distribution: India, Nepal, Bhutan, Sri Lanka

Recorded Localities and season: Dibrugarh, Sibsagar, Charaideo, Dehing Patkai, Summer,Autumn

Land use and land cover pattern: Forest clearing, gardens , parks, damp places

36. Commander, *Limenitis procris procris* Cramer

Sub family: Limenitinae

Genus: *Limenitis* (=*Moduza*)

Species: *procris*

Subspecies: *procris*

Distinguishing characters: Tawny, orange or red. Upper with discal white broad band. Coffee and brown coloured marginal lines with innermost line produced into arrowheads. In *Moduza*, discal vein straight. Fore cell end at origin v3, Fore cell closed, antenna longer than half costa. In *procris procris,* Upper fore end cell with large white spot and submarginal reddish brown band narrower than upper hind. Wing span: 60-75mm

Global Distribution: India, Myanmar, Bangladesh.

Recorded Localities and season: Kokrajhar, Kaziranga; monsoon

Land use and land cover pattern: Woodland, forest tracks, gardens

Forage plant: *Dahlia,* Marigold (*Tagetes*) *etc.*

37. Colour Sergeant, *Athyma nefte inara* Westwood

Sub family: Limenitinae

Genus: *Athyma*

Species: *nefti*

Sub-species: *inara*

Distinguishing Character: Upper black with tawny whitish marks. Cell streak broken into several yellow spots. Hind fore with discal and submarginal tawny bands. In *Athyma,* antenna longer than half costa. Fore upper end cell at origin V3. In *nefte inara,* female, fore apex with two rows of tawny marks in space 4,5,6 making apical bands. Markings broad. Sexual dimorphism present. In male, the inner apical band white, upperfore with cell streak dull yellow with a clearly detached white spot and upper subbasal band white. Wingspan: 68mm

Female

Male

Global Distribution: India, Bhutan

Recorded Localities and season: Dibrugarh; Dibrugarh University campus, Pre Monsoon

Land use and land cover pattern: Gardens, Forests.

Forage plant: Not recorded

38. Assam Unbroken Sergeant, *Athyma pravara* Fruhstorfer

Sub family: Limenitinae

Genus: *Athyma*

Species: *pravara*

Distinguishing Character: Black with upper marks white. In *Athyma,* antenna longer than half costa. In *pravara,* streak in cell is unbroken, basally narrow, a club at end and an elongated mark beyond, in discal region . Upper fore the spot out side cell in 3 is rounded and apex two rows of white marks. Marks large. Upper hind with post discal band of white marks. Protected under Schedule II of IWPA 1972.Wingspan: 60mm

Global Distribution: India, Bhutan

Recorded Localities and season: Kokrajhar, Dibrugarh University Campus; Post Monsoon

Land use and land cover pattern: Gardens, Forests.

Forage plant: Not recorded

39. Common Sergeant, *Athyma perius* Linnaeus

Sub family: Limenitinae

Genus: *Athyma*

Species: *perius*

Distinguishing Character: Black with upper marks white. In *Athyma perius,* fore cell open. In upperfore, streak in cell is broken into four, basally narrow. Under, yellow with white marks. Under hind with prominent black spots on the inner margin of the post discal white band. Wingspan: 60-70mm

Global Distribution: India, Bhutan, Burma

Recorded Localities and season: Jakai Biodiversity park,Dibrugarh, Dehing Patkai forest tracks ; Pre Monsoon

Land use and land cover pattern: Gardens, Forests.

Forage plant: Verbanaceae (*Lantana camera*), Asteraceae

40. Himalayan Sergeant, *Athyma opalina* Kollar

Sub family: Limenitinae

Genus: *Athyma*

Species: *opalina*

Distinguishing Character: Upper black with white marks. In *Athyma opalina*, upperfore, streak in cell twice divided , basally narrow and separate from the pointed, elongated spot. Upperfore spot in vs 1-3 directed towards termen. Wingspan: 55-70mm

Global Distribution: India, Bhutan, Burma

Recorded Localities and season: Jakai Biodiversity park, Dibrugarh, Dehing Patkai forest tracks ; Guwahati , Pre Monsoon

Land use and land cover pattern: Gardens, Forests.

Forage plant: Verbanaceae (*Lantana camera*), Asteraceae

41. Staff Sergeant, *Athyma selenophora* Kollar

Sub family: Limenitinae

Genus: *Athyma*

Species: *selenophora*

Distinguishing Character: Black with marks white. In *Athyma* fore cell open. In *selenophora,* upperfore, with streak in cell red broken into two. Apical white spots separate. Upper with white broad discal band. Under fore with cell streak white, broken into three. Wingspan: 55-75mm

Global Distribution: India, Bhutan, Burma, Bangladesh

Recorded Localities and season: Dehing Patkai forest tracks; Pre Monsoon

Land use and land cover pattern: Gardens, Forests.

Forage plant: Not recorded

42. Common Sailer, *Neptis hylus varmona* Moore

Sub family: Limenitinae

Genus: *Neptis*

Species: *hylus*

Sub-species: *varmona*

Distinguishing Character: Upper marks white. Streak in cell is broken into three fragments, first two cojoined and the discal spot end cell is triangular and outwardly sharply pointed. In *Neptis,* fore cell open. Palpi prominent and hairy. In *hylus,* line joining marks in 1 -3 point towards apex, and marks in 2 and 3 not in the line with marks in 5 and 6. Upper hind with one discal and one submarginal band of white patches. In, *hylus varmona,* upper part of discal band in upper fore is obliquely arranged and the costal streak is short. UNH white discal band with black margin. Bask under sun with wings open. Never found them closing their wings. Wingspan: 60mm

Global Distribution: India, Sri Lanka, Myanmar

Recorded Localities and season: Dibrugarh (University); Post Monsoon(Nov)

Land use and land cover pattern: Gardens, Forest edge

Forage plant: *Camelia sinensis*

43. Plain Sailer, *Neptis cartica* Moore

Sub family: Limenitinae

Genus: *Neptis*

Species: *cartica*

Distinguishing Character: Upper marks white or sullied. Streak in cell is broken into two fragments separated by a slit .The spot end cell is elongated and outwardly sharpened. In *Neptis,* fore cell open. Like in *clinia,* upper hind spots in post discal band are separated. Like in *clinia susruta,* the spot in 2 and 3 upperfore discal band are outwardly square or concave and well separated . Upperfore apical markings are well separated. Upperhind, submarginal discal band narrower and dirty white. Marks are not black edged. Bask under sun with wings open. Wingspan: 50-60mm

Global Distribution: India

Recorded Localities and season: Jakai Biodiversity Park, Dibrugarh . Pre Monsoon and Monsoon (May-July)

Land use and land cover pattern: Gardens, Forest edge

Forage plant: Not recorded

44. Yellow Sailer, *Neptis ananta* Moore

Sub family: Limenitinae

Genus: *Neptis*

Species: *ananta*

Distinguishing Character: Upper marks yellow. Streak in cell is broken

into three fragments and the one in end cell is triangular. In *Neptis,* fore cell open. Palpi prominent and hairy. In *ananta,* above markings orange yellow, below markings ochreous. Upper fore submarginal line broken and termen with an indent in the middle. Bask under sun with wings open. Fly like sailing. Wingspan: 60mm

Global Distribution: India, Sri Lanka, Myanmar

Recorded Localities and season: Dibrugarh ; Post Monsoon (Nov)

Land use and land cover pattern: Gardens, Tea plantations;

Forage plant: Not recorded

45. Common Lascar, *Pantoporia hordonia hordonia* Stoll

Sub family: Limenitinae

Genus: *Pantoporia*

Species: *hordonia*

Subspecies: *hordonia*

Distinguishing Character: Upper marks tawny against dark brown. Streak in cell has two notches facing costa. Upperfore apex with single tawny band. In *hordonia hordonia*, upperhind with two orange bands, the marginal being narrower with less zigzag border. Was found basking under sun with wings open. They move frequently from one place to another while basking. Wingspan: 40-45mm

Global Distribution: India, Sri Lanka, Myanmar

Recorded Localities and season: Jakai biodiversity park, Dibrugarh; Dibrugarh University campus. Monsoon (July)

Land use and land cover pattern: Gardens, Tea plantations;

Forage plant: Not recorded

46. Common Baron, *Euthalia aconthea sudhudana* Cramer

Sub family: Limenitinae

Genus: *Euthalia*

Species: *aconthea*

Subspecies: *sudhudana*

Distinguishing Character: Upper brown with deep olive green shade over brown at base. Cells open. In *Euthelia*, antenna with brown tipped gradual club. Upper hind with marginal band with dark brown or black spots. In *aconthea sudhudana*, pale patches beyond cell in spaces 2 to 6. In fore, costal white spot in female. Bask under sun with wings open . Protected under Schedule II of IWPA 1972. Wingspan: 60-70mm

Global Distribution: India, Sri Lanka, Myanmar, Bhutan

Recorded Localities and season: Dibrugarh (University), feeds on rotten vegetable and fruits.

Land use and land cover pattern: Gardens,

Forage plant: Not recorded ; Feed on rotten fruits.

47. Himalayan Common Baron, *Euthalia aconthea garuda* Moore

Sub family: Limenitinae

Genus: *Euthalia*

Species: *aconthea*

Subspecies: *garuda*

Distinguishing Character: Upper brown with white body underside . Cells open. In *Euthelia*, antenna with red tipped gradual club. Upper hind has marginal band of dark brown or black spots. In *aconthea garuda*, in both sides of fore, white prominent patches beyond cell in spaces 2 to 6 and smaller white marks(5) from space 5 to 7 in the form of a curve, the costal spot being the largest. Underwings paler with similar pattern of marks. Attracted to fruit bait. Proboscis is green in colour with orange tip. Bask under sun with wings open. Wingspan: 60-70mm

Global Distribution: India, Sri Lanka, Myanmar, Bhutan

Recorded Localities and season: Dibrugarh , feeds on rotten vegetable and fruits.

Land use and land cover pattern: Gardens

Forage plant: Not recorded: Feed on rotten fruits

48. Plain Earl, *Tanaecia jahnu jahnu* Moore

Sub family: Limenitinae

Genus: *Tanaecia*

Species: *jahnu jahnu*

Distinguishing Character: Upper dusky brown. In *Tanaecia,* fore cell closed, end cell discal line at origin V3 In *jahnu jahnu,* discal band is formed by crescents. No blue margin. Sub marginal row of darkbrown marks upperhind like bands and in upperfore like spots . From space 2 towards costa between discal and submarginal brand is whitish upperfore. Antenna club gradual. Fore apex produced yet truncate. Observed to be low flying over grass cover. Wingspan: 85-90mm

Recorded Localities and season: Jakai Biodiversity Park, Dibrugarh

Land use and land cover pattern: Gardens

Forage plant: Not recorded

49. Himalayan Grey Count, *Tanaecia lepidea lepidea* Butler

Sub family: Limenitinae

Genus: *Tanaecia*

Species: *lepidea*

Subspecies: *lepidea*

Distinguishing Characters: Upper dark brown with darker brown curved lines from costa downwards. Forewing apex curved downwards. Underfore pale with yellow patches and brown marks. Under hind is paler than underfore. I have found them in two sizes. The smaller one with brighter dark brown colour. In *Tanaecia,* fore cell closed, end cell discal line at origin V3. In *lepidea,* upper hind with broad shiny pale brown margin and upper fore narrow margin .Under fore brown discal line parallel to termen.

This is schedule II species under IWLP Act, 1972. Wingspan: 65 & 80mm.

Global Distribution: India, Bhutan, Myanmar, Bangladesh, SE Asia

Recorded Localities and season: Jakai Biodiversity Park, Dibrugarh; Post monsoon (Aug-Nov)

Land use and land cover pattern: Open gardens and parks

Forage plant: Not recorded

Subfamily5: Charaxinae (Rajahs, Nawabs)

50. Yellow Rajah, *Charaxes marmax marmax* Westwood

Sub family: Charaxinae

Genus: *Charaxes*

Species: *marmax*

Subspecies: *marmax*

Distinguishing Characters: Body and wings brown with yellow tinge. In *Charaxes*(Rajah), cells closed. Upper fore with dark brown mark broad at apex narrowed to tornus. Underfore pale yellow bands across wingspan at the level below discal band. In *marmax,* underfore cell has black bar with a spot . Under hind with submarginal white diffused spots. Under fore and hind with black lines between veins which give a pattern of curved lines when wings are closed. Proboscis colour yellow. Antennae black. Hind

wing toothed. Protected under schedule II of IWPA, 1972. Wingspan:90-120mm

Global Distribution: India(Assam), Nepal, Bhutan

Recorded Localities and season: Ultapani Reserve Forest, Kokrajhar; Monsoon

Land use and land cover pattern: Thick forest , come out for mud puddling in forest tracks.

Forage plant: Not recorded

51. Black Rajah, *Charaxes solon solon* Fabricius

Sub family: Charaxinae

Genus: *Charaxes*

Species: *solon solon*

Distinguishing Characters: Body and wings brown with darker tinge. In Charaxes(Rajah), cells closed.Under brown with dull bands. In *solon*, hind with a tail at 2 & 4. Upper dark brown with pale border . Underfore cell has black doublelined bar and several curved black lines. Under hind bears submarginal band with small black spots. Under fore and hind with black lines between veins which give a pattern of curved lines when wings are closed. Antennae black, club gradual. Males smaller. Wingspan:80-100mm

Global Distribution: India, Nepal, Bhutan

Recorded Localities and season: Dibrugarh; Monsoon

Land use and land cover pattern: Gardens, forest tracks.

Forage plant: Not recorded

52. Common Nawab, *Polyura athamas* Drury

Sub family: Charaxinae

Genus: *Polyura*

Species: *athamas*

Distinguishing Characters: Brown in colour. In Nawabs, H cell open. Hind wing with two sharp teeth, Wing or tail with green discal band. In *athamas*, UN hind a dark basal band passes through cell. Under hind with marginal black bordered white marks. Under fore with apical green spot apart from green discal band in both wings. The broad discal band is bordered by semilunar green marks. Wingspan: 75mm

Global Distribution: India, Bangladesh, Nepal, Bhutan

Recorded Localities and season: Near Kaziranga

Land use and land cover pattern: Forest and forest tracks, woodland, Mudpuddling over rotten organic materials

Forage plant: Not recorded

Subfamily 6: Nymphalinae (Wizard, Eggfly, Jester, Pansy)

53. Myanmarese Wizard, *Rhinopalpa polynice birmana* Fruhstorfer

Sub family: Nymphalinae

Genus: *Rhinopalpa*

Species: *polynice*

Subspecies: *birmana*

Distinguishing Characters: Body orange brown and wings chocolate

brown. Hind wing with two pointed teeth. Under with silvery marginal lines and submarginal row of white centered lunules. The lunules underhind are bordered by orange brown. In *Rhinopalpa,* cells open. Fore termen highly concave in middle and convex in 1&2. In *polynice birmana* underwing rich dark brown with very narrow curved silverlines. Protected under Schedule II of IWPA 1972 . Wingspan: 90-120mm

Global Distribution: India, Myanmar

Recorded Localities and season: Kokrajhar, forest trek; Monsoon

Land use and land cover pattern: Thick forest, come out for mud puddling in forest tracks

Forage plant: Not recorded

54. Chinese Lemon Pansy, *Junonia lemonias lemonias* Linnaeus

Sub family: Nymphalinae

Genus: *Junonia*

Species: *lemonias*

Subspecies: *lemonias*

Distinguishing Characters: Body brown. Apical margin of upper side of the wings are bordered by two rows of lemon yellow marks separated by dark brown scales . Upper fore contains lemon spots surrounding the eye at 2 and the anterior broad margin. In *Junonia,* antenna clubbed, about half costa in length. Cells open. In *lemonias lemonias,* upper each wing with one coffee and white centered yellow ringed large ocellus and another small ocellus. Underhind the cell base black.Wing span: 50-60mm

Global Distribution: India, Myanmar, Nepal, Bhutan, Andaman

Recorded Localities and season: Assam; Post monsoon

Land use land cover pattern: Mostly on herbs and shrubs

Forage plant: *Alternanthera* , *Zinnia* etc.

55. Grey Pansy, *Junonia atlites* Linnaeus

Sub family: Nymphalinae

Genus: *Junonia*

Species: *atlites*

Distinguishing Characters: Body and wings light brown. In *atlites,* upper side of both wings with submarginal row of ocelli with some orange and black centered. Cell open. Antenna clubbed and half costa in length. Wing span: 80mm

Global Distribution: India, Myanmar, Sri Lanka, Andaman

Recorded Localities and season: Assam; Post monsoon

Land use land cover pattern: Mostly on herbs and shrubs

Forage plant: *Tridax, Lantana camara,* etc.

56. Peacock Pansy, *Junonia almana* Linnaeus

Sub family: Nymphalinae

Genus: *Junonia*

Species: *almana*

Distinguishing Characters: Body and wings tawny red. Antenna clubbed and half costa in length. In *almana,* upper hind with single large peacock coloured centric dark brown ocellus with yellow border. Upper fore with three smaller ocelli, the apical two cojoined. Under pale brown. Underhind has a transverse white line . The line on under fore is not straight . Wing span: 60-65

Global Distribution: India, Myanmar, Sri Lanka, Andaman, Pakistan

Recorded Localities and season: Assam; Monsoon,Post monsoon

Land use and land cover pattern: Mostly on herbs and shrubs

57. Chocolate Pansy, *Precis iphita* Cramer

Sub family: Nymphalinae

Genus: *Precis (= Junonia)*

Species: *iphita*

Distinguishing Character: Upper brown with deep olive green shade over brown in thorax and dorsum of hind. In *Precis,* cells open. Antenna clubbed with brown tip. Fore vs 1B and 2 almost straight. Upper with two submarginal dark brown band . Rows of darkbrown centered pale Ocelli between veins in space between discal and submarginal brands. But no prominent ocelli as in other pansies. Upper fore apex produced but truncate and bears a white spot. No basal mark in space 1b. Bask under sun with wings open. Similar to chocolate pansy. Wingspan: 55-80mm

Global Distribution: India, Sri Lanka, Myanmar, Bhutan

Recorded Localities and season: River bankside in palm plants, Dibrugarh

Land use and land cover pattern: Gardens: Forage plant: Not recorded

58. Great Eggfly, *Hypolimnas bolina* Linnaeus (F)

Sub family: Nymphalinae

Genus: *Hypolimnas*

Species: *bolina*

Distinguishing Characters: Body black. Wings upper dark coffee brown and under lighter. Upper fore with blue shiny marks above cell near to costa. Upper side of both wings have marginal row of white curved lines. Under fore with white marginal and submarginal spots. In *Hypolimnas*, antenna clubbed, gradual. Fore apex round, truncate and termen concave just below. In *bolina,* underhind with white broad marginal band. In female, upper fore with row of white semi lunar marks followed by spots and upper hind with broad white margin. I have observed a female taking rest by hanging from the stem of a jasmine plant with wings closed for more than eight hours from 1.15 pm till 9.30 pm. When disturbed during rest till two hours, it slowly opened and closed the wings and went back to rest again. But after about two hours it was hanging like a dead one. Wingspan: 70mm

Global Distribution: India, Myanmar, Bangladesh, Andaman

Recorded Localities and season: Dibrugarh; Post monsoon (Nov)

Land use and land cover pattern: Open gardens and parks

Forage plant: Not recorded

59. Great Eggfly, *Hypolymnas bolina* Linnaeus (M)

Sub family: Nymphalinae

Genus: *Hypolimnas*

Species: *boina*

Distinguishing characters: Body black. Wings upper black and under lighter. Upper hind with white centered blue shiny oval marks one on each wing followed by row of submarginal spots. Upper fore with one apical and one central (three marks together) white mark. Upper with white marginal marks from apex to tornus. Wingspan-70mm

Global Distribution: India, Myanmar, Bangladesh, Andaman

Recorded Localities and season: Dibrugarh; Post monsoon (Aug-Nov)

Land use and land cover pattern: Open gardens and parks

Forage plant: Not recorded

60. Danaid Eggfly, *Hypolimnas misippus* Linnaeus (M)

Sub family: Nymphalinae

Genus: *Hypolimnas*

Species: *misippus*

Distinguishing characters: Body blackish blue. Antenna clubbed, fore apex round, truncate and termen concave just below Wings blackish on both sides. Wings with white oval discal marks . Fore wing apex with white marks . No marginal white spots like great egg fly. Protected under schedule I of IWPA 1972. Wing span-68-70mm

Global Distribution: India, Myanmar, Bangladesh, Andaman

Recorded Localities and season: Dibrugarh; Post monsoon (Aug)

Land use and land cover pattern: Open gardens and parks, rests in shade

among foliages with both wings opened.

Forage plant: Not recorded

61. Himalayan Autumn Leaf, *Doleschallia bisaltide indica* Moore

Sub family: Nymphalinae

Genus: *Doleschallia*

Species: *bisaltide indica*

Distinguishing characters: Oak leaf like in appearance. Cells open, apex broadly truncate. Upper tawny with black apex . The latter bears a tawny band outside cell. Antenna club gradual. Upper tawny with white banded black apex. Under dull rusty brown with white brand . Protected under Schedule II of IWPA,1972. Wing Span:75-80mm

Global Distribution: India, Myanmar, Bangladesh, Andaman

Recorded Localities and season: Bokakhat, Kaziranga, Post monsoon (Aug)

Land use and land cover pattern: Open gardens and parks, Forest treks

Forage plant: Not recorded

62. Jester, *Symbrenthia lilaea* khasiana Moore

Sub family: Nymphalinae

Genus: *Symbrenthia*

Species: *lilaea*

Distinguishing characters: Upper dark brown with tawny bands. Upper fore cell streak with two constrictions , streak basally narrow. Upperfore with single apical spot followed by an oblique broad bar. The hind produced to form a short tail at 3. Upper hind with broad discal and post discal bars. Under similar with fore and bears a long cell bar. Observed foraging close to ground. The species could not be identified. Wing Span: 45-55mm

Global Distribution: India, Myanmar, Bangladesh, Andaman

Recorded Localities and season: Jakai, Dibrugarh, Spring (May)

Land use and land cover pattern: Open gardens and parks, Forest treks

Forage plant: Not recorded

63. Common Jester, *Symbrenthia hippoclus* Cramer

Sub family: Nymphalinae

Genus: *Symbrenthia*

Species: *hippoclus*

Distinguishing Character: Upper marks tawny against dark brown. Streak in cell has two notches facing costa. In *Symbrenthia,* upperfore apex with one broad and one narrow tawny band. The hind between V2-V3 is produced. In *hippoclus khasiana,* under hind a dark brown line from near base V1 across cell to mid costa. Was found basking under sun with wings open. Wingspan: 45-55mm

Global Distribution: India, Nepal, Myanmar, Bhutan

Recorded Localities and season: Jakai biodiversity park, Dibrugarh; Dehing Patkai forest trek Monsoon (July)

Land use and land cover pattern: Forest treks, Gardens

Forage plant: Not recorded

64. Painted Lady, *Vanessa cardui* Linnaeus

Sub family: Nymphalinae

Genus: *Vanessa*

Species: *cardui*

Distinguishing characters: Upper buff, marked with black. Fore apex black with white marks. Discal area redish. Under cryptically marked with different colours. Upper fore submarginal prominent black spots. In *Vanessa,* antenna longer than half costa, cells closed. In *cardui,* fore red area with black spots. Base golden brown. Antennal club with yellow tip. Wing Span:60:70mm

Global Distribution: India, Myanmar, Bangladesh, Andaman

Recorded Localities and season: Dibrugarh, March-April

Land use and land cover pattern: Open gardens and parks.

Forage plant: Not recorded

Subfamily 7: Cyrestinae (Maps)

65. Common Map, *Cyrestis thyodamas* Boisduval

Sub family: Cyrestinae

Genus: *Cyrestis*

Species: *thyodamas*

Subspecies: *thyodomas*

Distinguishing Character: White winged with wavy longitudinal irregular thin lines. Apex of hind cut off. Upper tornus with orange bordered black spot. In *Cyrepsis,* antenna with slender club. In *thyodomas thyodomas,* upper almost white. Upper fore with marginal black line and apex with blackish grey marks. Wingspan: 50-60mm

Global Distribution: India, Myanmar, Nepal, Bhutan, Bangladesh

Recorded Localities and season: Kaziranga; Dehing Patkai, Pre monsoon

Land use and land cover pattern: Woodland, near waterbodies.

Forage plant: Compositae

Subfamily8: Biblidinae (Castors)

66. Common Castor, *Ariadne merione assama* Evans

Subfamily: Biblidinae

Genus: *Ariadne*

Species: *merione assama*

Distinguishing Characters: Body and wings upper rusty brown. Heavily waved double undulating lines. Redish marks between the wavy lines and marginal border. Under dull in colour. In *Ariadne*, antennae slender clubless. Fore apex truncate. In *merione assama* , the discal line beyond cell is straight. Wing span: 66mm

Global Distribution: India, Myanmar

Recorded Localities and season: Through out Assam; Pre monsoon, monsoon

Land use and land cover pattern: Cultivated lands, Gardens , forests tracks, woodland

Forage plant: *Ricinus Communis*

67. Angled Castor, *Ariadne pallidior* Frusthorfer

Subfamily: Biblidinae

Genus: *Ariadne*

Species: *pallidior*

Distinguishing Characters: Upper ochreus brown with regularly arranged black lines . Antenna slender, brown tipped, narrowly clubbed. Fore apex straight. In *pallidior*, upperhind veins 5 and 6 not white and discal lines run to costa. White distinct spot in fore apex near costa. Termen wavy. Under dull brown. Wing span: 71mm

Global Distribution: India, Myanmar

Recorded Localities and season: Through out Assam; Pre monsoon, monsoon

Land use and land cover pattern: Cultivated lands, Garden, open areas .

Forage plant: *Ricinus Communis*

Chocolate albatross, *Appias lyncida*

Family: Pieridae

Dipsikha Bora

Pierids are small or medium sized white or yellow coloured butterflies found in all types of habitats from grasses to deciduous and evergreen forests. Grass yellows and emigrants are very commonly found in gardens. Some can be very colourful and exhibit seasonal plasticity of colour and patterns. There may be different forms of dry and wet season. Mostly underside of the hindwings show cryptic colouration.

Pierids have broad forewings. Hv1a is present. The thorax is of medium size except for the fast fliers. Wings mostly have basal folds. All the three pairs of legs are normal. They mostly fly close to the ground level. Some are slow fliers while others may be moderate to swift fliers. Emigrants, cabbage butterflies and albatrosses are studied for their migratory behaviour for adapting to seasonal changes of environmental factors but mostly in USA, Australia and Europe. Many of them are sun loving, appearing during sunny period of the day and hence most noticeable. They follow a diurnal rhythm for activity. They may be seen mud puddling alongside butterflies of other families. During rest, wings are always kept folded up at right angle.

They usually lay eggs singly, though some lay eggs in batches. Larvae feed on plants belonging to Cruciferae, Fabaceae, Caesalpiniaceae, Salvadoraceae, Euphorbiaceae, Geraniciaceae, Viscaceae and because of such polyphagy can have wide geographic distribution. Adults nectaring in flowers of grasses and seasonal flowers planted in gardens is a common sight in Assam. Larvae are of uniform width from thorax and abdomen. Pierids like jezebels, grass yellow pupate in groups in cocoons. The cocoons may be securely tied to branches using silk girdles. The Pierids are identified mostly by their colour patterns including markings (spots, brands) venation pattern, wing shape and sometimes presence or absence of characters exhibiting sexual dimorphism like hair pencils and wing scale colours of red, yellow, tawny, orange, black and white. About 1164 Pierid butterflies are identified in the world out of which 114 species have been identified in India. The Pierids in this book are represented by two subfamilies, Pierinae including the whites and Coliadinae including the yellows

Subfamily: Pierinae (Gulls, Jezebels, Albatross)

1. Psyche, *Leptosia nina nina* Fabricius

Subfamily: Pierinae

Genus: *Leptosia*

Species: *nina nina*

Distinguishing Characters: Body and wings white. In *Leptosia,* in hind the cell angle in apex is acute and discocellular vein is concave. V8 is approximately of same length as cell. In *nina nina,* upperfore with a black bordered apex and a prominent black spot. Underside with faint green marks. I have observed them hovering after herbaceous flowers during sunny noons. They exhibit restless, slow movement for nectaring. At times they fly fast with a jerking movement. Wingspan: 35mm

Global Distribution: India, Nepal, Bhutan, Parts of Oriental region

Recorded Localities and season: Assam ; Spring, Monsoon, Postmonsoon.

Land use land cover pattern: Mostly on herbs in shaded places, vegetable garden, close to ground

Forage plant: Flowers of common herbs of Assam like *Ageratum* .

2. Large Cabbage White, *Pieris brassicae* Linnaeus

Subfamily: Pierinae

Genus: *Pieris*

Species: *brassicae*

Distinguishing Characters: Upper fore white with black apex and prominent black spot. Wings with black veins. In fore V8 absent, in hind V7 absent. In *brassicae*, no dark marks on apical margin underside. Upperfore with two dark spots between space 1 and 3. Under hind is pale yellow with a small black mark end of costa. Wingspan: 75mm

Global Distribution: India, Himalaya

Recorded Localities and season: All throughout Assam from Sadiya to Dhubri; all seasons

Land use land cover pattern: Common in open areas on herbs and shrubs near gardens, cultivated lands

Forage plant: *Vitex, Scutellaria, Brassica*

3. Indian Cabbage White, *Pieris canidia indica* Evans

Asian cabbage white, *Pieris canidia* Linnaeus 1768

Subfamily: Pierinae

Genus: *Pieris*

Species: *canidia*

Sub species: *indica*

Distinguishing Characters: Upper fore white with black apex bearing toothed inner margin. A black spot on forewing outside the cell is visible from both surfaces of the wing. In *canidia*, under hind with no black markings. Under hind, pale yellow dusting and base of costa is distinct yellow. Wingspan: 60 mm

Global Distribution: India, Himalaya

Recorded Localities and season: Assam; all seasons

Land use land cover pattern: Common in open areas on herbs and shrubs near gardens, cultivated lands

Forage plant: *Vitex, Scutellaria,* flowers of Brassicaceae

4. **Common Gull, *Cepora nerissa* Fabricius (F)**

Subfamily: Pierinae

Genus: *Cepora*

Species: *nerissa*

Distinguishing Characters: Body white with brown dusting. Under veins chocolate brown. In *Cepora*, in forewing vein 9 is absent. Forewing not produced. Upperhind with marginal row of marks having chocolate brown border. Upper cell bordered with light chocolate brown. In the species, *nerissa* under, veins are chocolate brown and yellow at base. Upperfore predominantly white with light chocolate brown apex narrowing towards tornus. I observed a female's restless wandering for an hour before oviposition on a stem above 2m from ground. Eggs are laid singly. Amplexiform coupling apparatus formed by overlapping of forewing over hindwing. Wingspan: 61mm.

Global Distribution: India (Sikkim, Assam), Burma

Recorded Localities and seasons: Zoological garden, Guwahati , Assam, Monsoon (July)

Land use land cover pattern : Shrubs and herbs in Parks, gardens;

Forage plant: Lantana, Flowers of herbs and shrubs

5. **Sylhet Common Albatross, *Appias albina darada* C&R Felder (Form II)**

Common albatross, *Apias albina* Boisduval

Subfamily: Pierinae

Genus: *Appias*

Species: *albina*

Subspecies: *darada*

Distinguishing characters: In *Appias*, apex of fore wing produced. Upper fore with black costa. In *albina darada*, under greenish yellow, upper

pale . Compound eye greenish. Underhind costa with prominent yellow base. Flies very fast. Wing span:60-75mm (Haribal,1992). The subspecies *darada* is a scheduled II species of Indian Wildlife Protection Act, 1972.

Global Distribution: India(Sikkim, Assam), Burma, Albana

Recorded Localities and season: Kaziranga

Land use land cover pattern: Forest, woodland, flower garden.

Forage plant: Not recorded

6. Western Striped Albatross, *Appias libythea* Fabricius (M)

Subfamily: Pierinae

Genus: *Appias*

Species: *libythea*

Distinguishing Characters: Body with black white hairs. In *Appias,* fore wing produced. In *libythea* (male), upper fore with black costa, termen with inwardly pointed black marks between veins making a wavy marginal

band. Upperhind margin black, no inward markings. Under white. Seen in large numbers. Protected under Schedule IV of IWPA 1972. Wingspan: 50mm.

Global Distribution: India, Sri Lanka, Myanmar, Bhutan, Pakistan, Nepal

Recorded Localities and season: Guwahati, Dibrugarh, monsoon (July) to winter(November).

Land use land cover pattern: Open areas, gardens (Botanical, zoological)

Forage plant: *Eupatorium* spp.

7. **Western Striped Albatross, *Appias libythea* Fabricius (F)**

   Subfamily: Pierinae

   Genus: *Appias*

   Species: *libythea*

Distinguishing Characters: Body black with white hairs. In *Appias,* fore wing produced. In *libythea* (female), fore with dark brown costa, cell and apex. Fore termen with dark brown border. No pale spot on dark border of fore. Hind with very faint brown marginal dusting. This is dry season form. Protected under Schedule IV of IWPA 1972. Wingspan: 50-60 mm.

Global Distribution: India, Sri Lanka, Myanmar, Bhutan, Pakistan, Nepal

Recorded Localities and season: Guwahati, Dibrugarh, Post monsoon, November.

Land use land cover pattern: Open areas, gardens (Botanical, zoological)

Forage plant: *Eupatorium* spp.

8. **Eastern Striped Albatross, *Appias olferna* Swinhoe (F) [Form I]**

   Subfamily: Pierinae

   Genus: *Appias*

   Species: *olferna*

Distinguishing Characters: In *Appias*, fore wing produced. In *olferna* female, upper white. upperfore, apex and termen black with white marks. In upper fore, anterior part of cell black. The black discocellular mark extends to join black border of termen. The marginal black marks upperhind extend narrowly along veins. This is dry season form.Wingspan: 50-60 mm.

Recorded Localities and season: Dibrugarh, Winter (December).

Land use land cover pattern: Open areas, gardens (Botanical, zoological)

Forage plant: *Eupatorium* spp.

9. Eastern Striped Albatross, *Appias olferna* Swinhoe (F)[Form II]

Subfamily: Pierinae

Genus: *Appias*

Species: *olferna*

Distinguishing Characters: Upper brown with white marks, under light brown. Upper marginal row of white spots. Under veins distinct brown. Under fore half basal and underhind costal and dorsum with yellow tinge. Wingspan: 45-50 mm.

Global Distribution: India, Sri Lanka, Myanmar.

Recorded Localities and season: Dibrugarh; Post monsoon (November).

Land use land cover pattern: Shrubs and herbs in parks and gardens

Forage plant: *Eupatorium* spp, *Ageratum, Biden* etc.

10. Eastern Striped Albatross, *Appias olferna* Swinhoe (M)

Subfamily: Pierinae

Genus: *Appias*

Species: *olferna*

Distinguishing Characters: Body with black white hairs. In *Appias*, fore wing pointed. Upper white. In *olferna* (Male), upper fore apex with three elongated white marks and the black border on termen is heavily indented along veins. Cell without any spot. Under hind pale yellow, veins black or dark brown. Underfore veins dark at apex, v8 , underhind v7,8, a narrow curved dark band from which arise dark lines running to termen. I observed them to be very active in sunny noon for nectaring. Sexual dimorphism present. Wingspan: 55-60 mm.

Global Distribution: India, Burma.

Recorded Localities and season: Guwahati, Dibrugarh; Winter (December).

Land use land cover pattern: Shrubs and herbs in parks and gardens

Forage plant: *Eupatorium* spp, *Ageratum, Biden* etc.

11. Chocolate Albatross, *Appias lyncida eleonara* Boisduval

Subfamily: Pierinae

Genus: *Appias*

Species: *lyncida*

Subspecies: *eleonara*

Distinguishing Characters: Forewing produced. In *lyncida eleonara*, under hind yellow with dark chocolate broad margin on termen. Under fore with black costa and three yellow elongated spots on black apex. Underfore black termen dentate on white. Upperfore black apex with three white spots. Found mudpuddling with other *Appias* species.Wingspan: 55-70 mm.

Global Distribution: India (Sikkim, Assam, South India, Arunachal Pradesh), Bhutan, Myanmar

Recorded Localities and season: Jakai Biodiversity Park, Dibrugarh, Kaziranga; Monsoon (July)

Land use land cover pattern: Forests and forest openings in any height

Forage plant: *Eupatorium* spp.

12. East Himalayan Yellow Orange Tip, *Ixius pyrene familiaris* Butler

Old Name: Yellow orangetip, *Ixius pyrene* Linnaeus

Subfamily: Pierinae

Genus: *Ixius*

Species: *pyrene*

Subspecies : *familiaris*

Distinguishing Characters: Antenna with yellow tip. In *Ixias,* v6 and 7 forked like many others. In *pyrene,* upperside of the wing is yellow with orange marks in the apex bordered by black boundary. Hind upper yellow with black marginal band. Underside is yellow. Sexual dimorphism in wing pattern is reported. Wingspan: 60-70 mm.

Global Distribution: India, Taiwan, Myanmar

Recorded Localities and season: Kokrajhar, Tezpur, Dibrugarh, Arunachal pradesh; Monsoon, post monsoon

Land use land cover pattern: Open sunlight areas in forest tracks, Parks

Forage plant: *Sesamum* spp., *Ixora, Jasminum*

13. White Orange Tip, *Ixias marianne* Cramer (M)

Subfamily: Pierinae

Genus: *Ixius*

Species: *mariane*

Distinguishing characters: Antenna white tipped. In *mariane,* upper fore white with broad black borders and orange elongated patches in apex. Orange marks extend below v2. Upperfore basal dusted with black. Females have black spots in the orange patches. Upper hind also is white with black marginal border. Underhind rusty brown . Wingspan:50-55mm

Global distribution: India

Recorded Localities and season: Kaziranga; Monsoon, post monsoon

Land use land cover pattern: Woodland, forest tracks

Forage plant: *Sesamum* spp., *Ixora, Jasminum*

14. Oriental Great Orange Tip, *Hebomoea glaucippe glaucippe* Linnaeus

Subfamily: Pierinae

Genus: *Hebomoea*

Species: *glaucippe*

Subspecies: *glaucippe*

Distinguishing Characters: Body white. In *Hebomoea,* hind with v 6 & 7 well separated at origin. In *glaucippe glaucippe* , forewing upper with orange apex bound by broad apical black border. The orange tip bears black markings between veins. The underside is pale yellow with rusty spots. I have observed them mudpuddling with other species of butterflies particularly the Papilioniids. They do nectaring steadily for seconds even during noon, but move frequently and rapidly between flowers before steady nectaring. Wingspan: 95-100 mm.

Global Distribution: India (Assam, Arunachal Pradesh), Myanmar, Taiwan, Nepal, Burma

Recorded Localities and seasons: Forest treks, Kokrajhar, Udalguri, Tezpur, Dibrugarh, Namsai, Arunachal Pradesh(Monsoon, Post Monsoon)

Land use land cover pattern: Open sunlight areas near canopies and moist areas in forest, Parks.

Forage plant: *Sesamum* spp., *Lantana camara, Ixora, Jasminum, Hibiscus*

15. Common Jezebel, *Delias eucharis* Drury

Subfamily: Pierinae

Genus: *Delias*

Species: *eucharis*

Distinguishing characters: In *Delias,* club of antenna slender. In *eucharis,* underfore costa and apical margin black, cell white with black stripes. Underhind basal yellow with red marginal spots edged with white between veins. They are known to lay eggs in batches. Wingspan: 81mm

Global Distribution: India, Sri Lanka, Myanmar, Nepal, Bhutan

Recorded Localities and season: Kaziranga

Land use land cover pattern and season: Forest edges, Wood land and flower gardens

Forage plant: *Lantana camara, Tridax, Magnifera indica*

16. Red Base Jezebel, *Delias pasithoe* Linnaeus

Subfamily: Pierinae

Genus: *Delias*

Species: *pasithoe* (= *aglaia*)

Distinguishing characters: In *Delias*, V11 absent. Antennal club slender. Upperhind with red basal and yellow tornal patch. Base of costa has a bright yellow mark. Upperfore with basal greyish dusting forming streaks and apical black with greyish marginal marks. In *aglaia*, UNH with red basal markings extending from costa to dorsum, bordered by black. Extensive yellow patches underhind with black marginal border along termen. Under fore black with grey patches. The larva gives out silken thread to tie the dark brown pupa. I have observed it flying to the height of about 20ft above the tree canopies. A fast flier. Wing expanse: 82 mm

Global Distribution: India(Sikkim, Assam, S. India) Nepal, Myanmar

Recorded Localities and season: Kaziranga, Dibrugarh University Campus

Land use land cover pattern and season: Forest tracks and flower gardens

Forage plant: *Callistemon* spp.

17. Red Spot Jezebel, *Delias descombesi descombesi* Boisduval

Synonym: Vietnamese red spot Jezebel

Subfamily: Pierinae

Genus: *Delias*

Species: *descombesi*

Subspecies: *descombesi*

Distinguishing characters: Under hind yellow with black margin and yellow marginal spots. In *descombesi descombesi* , underhind with basal red spot along half costa. Under fore black with greyish streaks and yellow-white submarginal streaks and spots.

Global Distribution: India, Myanmar

Recorded Localities and season: Kaziranga

Land use land cover pattern and season: Forest tracks, woodland and flower gardens

Forage plant: Forest flowers, *Callistemon*

Subfamily 2 : Coliadinae (Yellows)

18. Oriental Lemon Emigrant, *Catopsilia pomona pomona* Fabricius (Form I)

Subfamily: Coliadinae

Genus: *Catopsilia*

Species: *pomona*

Subspecies: *pomona*

Distinguishing Characters: Body greenish yellow, wing with yellow green shades. In *Catopsilia*, fore, apex lies between v7 & 8. Bases of all the four wings yellow. In *pomona*, the underhind is unmarked,without strigae, but a bright yellow patch underhind attached to upper cell base. Upper fore costa black. Antenna black. Wingspan: 80 mm.

Global Distribution: India, Myanmar, south east Asia, Australia

Recorded Localities and season: Zoological garden, Guwahati, Assam, July

Land use land cover pattern: upper canopies of shrubs and trees in Parks, Gardens, forests.

Forage plant: *Cassia* spp., *Lantana camara*

19. Oriental Lemon Emigrant, *Catopsilia pomona* Fabricius (Form II)

Subfamily: Coliadinae

Genus: *Catopsilia*

Species: *pomona*

Distinguishing Characters: Yellow. Upperfore with black apex and narrow marginal border. In *Catopsilia*, fore, apex lies between v7 & 8. Under light yellow with brown veins and reddish brown ringed silver spots. In *pomona*, no strigae are present . Wing expanse: 70 mm.

Global Distribution: India, south east Asia

Recorded Localities and seasons: Dibrugarh (University campus); Premonsoon (March)

Land use land cover pattern: Upper canopies of shrubs and trees in Parks, Gardens.

Forage plant: *Cassia* spp., *Mussaenda* , *Catharanthus, Tagetes, Cineraria*

20. Mottled Emigrant, *Catopsilia pyranthe minna* Herbst (Form I)

Subfamily: Coliadinae

Genus: *Catopsilia*

Species: *pyranthe*

Subspecies: *minna*

Distinguishing Characters: Body pale lemon yellowish white. In *pyranthe minna,* upper appearance chalky greenish. Bases of all four wings are pale yellow. Upper fore with dark brown apex extending narrowly in termen. UPF, a black spot at cell-end . The underhind is unmarked, with colourless strigae. It was found in injured state. Wingspan: 60-80 mm.

Global Distribution: India, Myanmar, south east Asia, Australia, South Africa

Recorded Localities and season: Zoological Garden, Guwahati , Assam; monsoon (July)

Land use land cover pattern: upper canopies of shrubs and trees in Parks, Gardens, road sides, forests.

Forage plant: *Cassia* spp., *Lantana camara, Mussaenda* sp

21. Oriental Mottled Emigrant, *Catopsilia pyranthe* Linnaeus (Form II)

Subfamily: Coliadinae

Genus: *Catopsilia*

Species: *pyranthe*

Distinguishing Characters: Body pale greenish white. Forewing apex black and extends along termen. Males have narrow boundary. Upper fore with cell-end bearing a black spot. Under hind and fore with red strigae. Under fore with one red edged silverspot endcell and underhind with two red edged silver spot at base v5&6. Underhind both wings bear submarginal row of red small spots. Wingspan: 65 mm.

Global Distribution: India.

Recorded Localities and season: Dibrugarh (University campus);

Premonsoon (March)

Land use land cover pattern: Upper canopies of shrubs and trees in Parks, Gardens, forests.

Forage plant: *Cassia* spp., *Lantana camara, Catharanthus*

22. Mottled Emigrant, *Catopsilia pyranthe* Linnaeus (Form III)

Subfamily: Coliadinae

Genus: *Catopsilia*

Species: *pyranthe*

Distinguishing Characters: Body pale lemon yellow. Appearance greenish. Upper fore with dark brown apex. In *pyranthe,* under cell-end with single reddish brown ringed silver spot . No red striations. Antenna reddish brown. Upper fore costa dark brown. Strigae extensive. Found while mudpuddling. Wingspan: 70 mm.

Global Distribution: India, south east Asia

Recorded Localities and seasons: Kaziranga, Dibrugarh, Monsoon, postmonsoon

Land use land cover pattern: Upper canopies of shrubs and trees in Parks, Gardens, forests. Mud puddle on damp place

Forage plant: *Cassia* spp., *Lantana camara, Mussaenda* sp.

23. Oriental Mottled Emigrant, *Catopsilia pyranthe pyranthe* Linnaeus (Form IV)

Distinguishing Characters: Body pale greenish yellow. Upper fore with black margin. Under with red strigae. Under, single spot at cell-end. Under veins yellow . I have observed one exhibiting the phenomenon of mimicry

by taking rest in a *Mussaenda* plant during noon. During rest, it was hanging down with wings folded.Wingspan: 65 -70 mm.

Global Distribution: India.

Recorded Localities and season: Dibrugarh ; Postmonsoon (Nov)

Land use land cover pattern: Upper canopies of shrubs and trees in Parks, Gardens, forests.

Forage plant: *Cassia* spp., *Lantana camara, Mussaenda*

24. Tree Yellow, *Gandaca harina assamica* Moore

Subfamily: Coliadinae

Genus: *Gandaca*

Species: *harina*

Subspecies: *assamica*

Distinguishing Characters: Body pale greenish yellow. The vein of 12 at base under fore is swollen , apex between v7 to v9 and discocellular vein concave. In *harina* , no marking except for a very thin black apex and an apical margin upperfore. In *harina assamica* ,the apical margin never below v3. Found visiting damp place in a rainy day. Female whitish. Wingspan: 45-50 mm.

Global Distribution: India.

Recorded Localities and season: Jakai Biodiversity Park, Dibrugarh ; Monsoon (July)

Land use land cover pattern: Parks, Gardens, forests clearings.

Forage plant: Not recorded

25. Three Spot Grass Yellow, *Eurema blanda* Boisduval

Subfamily: Coliadinae

Genus: *Eurema*

Species: *blanda*

Distinguishing Characters: Body and wings bright yellow. Forewing

upper with black incurved apex border. In *blanda*, the under fore with three spots in cell and a discocellular mark. The underside is yellow with rusty spots and a rectangular shaped black mark on the apex. Wingspan: 40-50 mm.

Global Distribution: India(Assam, Sikkim, South India), Myanmar, Taiwan, Bhutan, Sri Lanka

Recorded Localities and season: Assam, All seasons

Land use land cover pattern: Flowers in lower herbs, shrubs, upper canopies in Parks, Gardens.

Forage plant: *Lantana, Caesalpinia, Ageratum , Cassia, Parthenium etc .*

26. One Spot Grass Yellow, *Eurema andersonii* Moore

Subfamily: Coliadinae

Genus: *Eurema*

Species: *andersonii*

Distinguishing Characters: Body and wings bright yellow. Forewing

upper with black incurved apex border between v2 and v4. The under fore apical area with two dark brown streaks. Underhind black marginal spots. In *andersonii,* underfore one zigzag spot in cell. Pupa dark brown, has two white stripes underside. Under dusting is rarely seen. In form II, only one fain zigzag mark is observed in cell underfore. The insect was found using proboscis in a cow dung plot. Protected under Schedule II of IWPA,1972. Wingspan: 45 mm.

Global Distribution: India

Recorded Localities and season: Guwahati, Dibrugarh ; Summer

Land use land cover pattern: Flowers in lower herbs, shrubs, upper canopies in Parks, Gardens.

Forage plant: *Cassia, Celosia.*

27. Common Grass Yellow, *Eurema hecabe* Linnaeus

Subfamily: Coliadinae

Genus: *Eurema*

Species: *hecabe*

Distinguishing Characters: Body and wings yellow. Forewing upper with

black border from apex downwards which is incurved twice. Upper hind with somewhat uniform black border . In *hecabe*, underfore, the costal margin black, cell with two spots. Underhind with various small brown markings. Wingspan: 40-50 mm.

Global Distribution: India, Myanmar, Taiwan, Burma, Sri Lanka, Andamans

Recorded Localities and season: Assam; All seasons

Land use land cover pattern: Flowers in lower herbs, shrubs, upper canopies in Parks, Gardens.

Forage plant: *Lantana, Caesalpinia, Ageratum, Cassia, Celosia, Bauhinia* etc.

28. Small Grass Yellow, *Eurema brigitta* Stoll

Subfamily: Coliadinae

Genus: *Eurema*

Species: *brigitta*

Distinguishing Characters: Body and wings yellow. Forewing upper with

black excaveted apex and black termen. In *brigitta,* underfore, the costal margin black, cell with discocellular mark only . Under fore and hind with very small marginal spots. Wingspan: 40 mm.

Global Distribution: India, Myanmar, Sri Lanka

Recorded Localities and season: Assam; All seasons

Land use land cover pattern: Flowers in lower herbs, shrubs, upper canopies in Parks, Gardens.

Forage plant: *Lantana, Caesalpinia, Ageratum, Cassia, Celosia* etc.

Family: Lycaenidae

P. Chetia, DS Bora and A. Mech

Lycaenids are mostly small butterflies with conspicuous colour on the wings. They are popularly called as 'The blues', as many of them have blue or purple scales or sheens on the upperside of the wings. Many of the species show sexual dimorphism as regard the colour pattern. The family Lycaenidae is divided into subfamilies, Poritiinae, Miletinae, Curetinae, Theclinae, Lycaeninae, Polyommatinae, Riodininae. Lycaenids like grass blues , hedge blues and ceruleans are found visiting flowers of grasses, shrubs and the lineblues, hedge blues, are found attracted to human sweat and animal droppings. They are sun loving and many are found to bask on low lying grasses (Poaceae). These butterflies are identified by venation, colour patterns, spots and marks on both sides of the wings, tail length, antenna pattern, tibial strength, presence or absence and position of androconia and genitalia. Some of the butterflies bear tail on the hind wing while others are tailless. The tailed ones are often accompanied by brightly coloured ocelli and lobe. The colourful visual patterns of Lycaenids are reported to be due to presence of multilayered scales. The multilayering may be associated with thermoregulation strategies of the butterflies.

Eggs are spherical in shape with flattened ends and not smooth. The caterpillars can retract their head under a sclerotized prothoracic shield and have a comparatively thicker integument. They (Lycaeninae) have honey glands on the seventh segment that attract ants. A single celled epidermal gland called as pore cupola organ secrete substances to pacify aggression among ants. And thus the caterpillars maintain mutualistic relations with ants. They also exhibit predatory behaviour by feeding on other small insects like scales, aphids.

More than 6000 species of Lycaenids are identified in the globe out of which about 450 species are described in India and in this book we are reporting 38 species including sex based forms. They represent the subfamilies, Poritiinae, Miletinae, Theclinae, Lycaeninae, Polyommatinae and Riodininae. Theclinae includes the strong blues equipped with massive thorax represented by oak blues, tits, royals. Oak blues are bigger in size (34-65mm) as compared to other lycaenids(18-42mm). The Royals which are forest dwellers, may prefer carnivore droppings. Lycaeninae includes the weak blues, saphires and quakers. Polyommatinae includes the weak blues like grass blues, hedge blues, line blues, cerrulians and Riodinidae includes *Punchinello*. Inclusion of Riodinidae is controversial due to presence of nymphalid like forelegs in males.

Subfamily 1: Poritiinae (Jems)

1. Himalayan Common Jem, *Poritina hewitsoni* Moore

Sub family: *Poritiinae*

Genus: *Poritina*

Species: *hewitsoni hewitsoni*

Distinguishing characters: Tailless blue. In Paritiinae, Indian form, no tuft of black hair or brand on fore . In *Poritina,* upper black with broad shiny greenish blue marks outside cell incompletely divided by a black streak in 1b. In *hewitsoni,* upper with submarginal blue spots between veins. Upper fore with black apex and greenish blue bar endcell. In *hewitsoni hewitsoni,* upper hind cell black. Under brown with marks. The species is protected under Scheduled II of IWP Act, 1972 Wing span: 30-38mm

Global Distribution: India, Bhutan, Myanmar

Recorded Localities and season: Dibrugarh; Post-monsoon (November)

Land use and land cover pattern: Gardens near forest

Forage plant: Not recorded

Subfamily 2: Miletinae (Apeflies, Mottles)

2. Oriental Apefly, *Spalgis epeus epeus* Westwood

Sub family: Miletinae

Genus: *Spalgis*

Species: *epeus*

Sub- species: *epeus*

Distinguishing characters: Tailless blue. Upper brown with white margin, antennal club thick In *Spalgis,*under very pale brown with numerous

fine dark brown wavy lines. Under hind with darkbrown spot in cell. In *epeus epeus*, a well defined discal white patch is present on upperfore. Occasionally spreads the wings open during basking. Move restlessly before oviposition. Not common. Wing span: 20-30mm

Global Distribution: India, Bhutan, Myanmar

Recorded Localities and season: Dibrugarh, riverside; Post-monsoon (August)

Land use and land cover pattern: Bamboo hedge, gardens

Forage plant: Not recorded

Subfamily3: Theclinae (Tits, Flash, Oak Blues etc.)

3. Sikkim Common Tit, *Hypolycaena erylus himavantus* Fruhstorfer (M)

Sub family: Theclinae

Genus: *Hypolycaena*

Species: *erylus*

Sub- species: *himavantus*

Distinguishing characters: Tailed blue. In *Hypolycaenia*, the hind two

tails are present, the one in v1b is longer than the one in v2. Veins 5,6 and 7 are separate in forewing. In *erylus himavantus*, . upper fore dark shining blue with black apex narrowed down to tornus. Upper fore with a diffused black discal patch. Upper hind tornus with white edged orange crowned black spot. Underhind greyish brown with double line end cell. Under fore discal band continuous and underhind fragmented. Wing span: 32-36mm

Global Distribution: India, Bhutan, Myanmar

Recorded Localities and season: Guwahati suburbs,Kokrajhar, Kaziranga; monsoon

Land use and land cover pattern: forest tracks, gardens

Forage plant: Not recorded

4. Sikkim Common Tit, *Hypolycaena erylus himavantus* Fruhstorfer (F)

Sub family: Theclinae

Genus: *Hypolycaena*

Species: *erylus*

Sub species: *himavantus*

Distinguishing characters: Tailed blue. In *Hypolycaenia,* the hind two tails are present, the one in v1b is longer than the one in v2. The tails are upper black and lower white. In *erylus himavantus* female . upper greenish brown. Upper, both wings have dark discal band from subcostal to dorsum. Upperhind with white post discal band becoming indistinct from space4. Upper hind with prominent white edged black spot in marginal space2 . Under characters are same as in male. Wing span: 32-36mm

5. Oriental orchid Tit, *Chliaria othona othona* Hewitson

Sub family: Theclinae

Genus: *Chliaria*

Species: *othona*

Sub- species: *othona*

Distinguishing characters: Tailed. Upper blue. Under pale brown with blue tinge. In *Chliaria,* tail at v1b and v2 in hind. Underfore, black costal spot above midcell. Underfore with apex pale orange narrowing down to tornus and outside cell black edged orange patches. Under hind with black edged orange patches and streaks. Underhind with bright orange crowned black spot towards tornus. In *othona,* under hind a prominent black spot in space7 towards base. Under fore discal band broken at 4. The pair of cell-end lines black. A single blue bar in tornus between orange crowned spots. Protected under Schedule I of IWPA, 1972.Wing span: 24-27mm

Global Distribution: India, Bhutan

Recorded Localities and season: Gibbon Wild Life Sanctuary; Monsoon

Land use and land cover pattern: forest tracks and edge, found attracted to human sweat.

Forage plant: Not recorded

6. Fluffy Tit, *Zeltus amasa amasa* Fabricius (M)

Sub family: Theclinae

Genus: *Zeltus*

Species: *amasa*

Sub- species: *amasa*

Distinguishing characters: Tailed blue. In *Zeltus,* tail long at v1, shorter at

v2 and fluffy. In *amasa*, underhind tornus with black spots not bounded by orange. Under, both wings with distinct ochreous double lines in cell-end. Under fore ochreous brown apically with darker marginal rows of marks, submarginal and discal dark ochreous bands from costa to dorsum. Under hind upper area with ochreous brown tinge and the rest is white. Underhind with prominent costal spot at 7 near base. In male upper pale blue tinge at base, dark blue apex. Wing span:28-32

Global Distribution: India, Bhutan, Myanmar

Recorded Localities and season: Kaziranga; Monsoon

Land use and land cover pattern: forest tracks, gardens

Forage plant: Not recorded

7. Fluffy Tit, *Zeltus amasa amasa* Fabricius (F)

Sub family: Theclinae

Genus: *Zeltus*

Species: *amasa*

Distinguishing characters: Tailed blue. Hindwing with two white fluffy tails, the one in v1b is longer than the one in v2. Upper dark brown with bluish sheen, basal more bluish. In female *amasa*, upperhind dorsum and marginal border white. A prominent white edged black spot upperhind in space 2 on a white area. Upperhind tornal spot white, submarginal white mark in 1-3 and hind margin completely white. Antennal club yellow tipped. Underhind characters are same as those of male. Wing span:28-32

Global Distribution: India

Recorded Localities and season: Kaziranga; Monsoon

Land use and land cover pattern: forest clearings

8. Yamfly, *Loxura atymnus* Cramer

Sub family: Theclinae

Genus: *Loxura*

Species: *atymnus*

Distinguishing characters: Under ochreous with darkbrown spots toward base and dorsum of hind. In *Loxura*, the hind produced till v2, extended to form white tipped tail. In *atymnus*, upperfore with black apex. Under a post discal band, across the wings. Sexual dimorphism is present in spot pattern towards dorsum underhind. These strong fliers were observed moving high up in the air in the copulating stage. Wing span:36-40mm

Global Distribution: India, Nepal, Bhutan, Myanmer.

Recorded Localities and season: Dibrugarh University Campus; Post Monsoon(August)

Land use and land cover pattern: Botanical Garden

9. Slate Flash, *Rapala manea schistacea* Moore

Sub family: Theclinae

Genus: *Rapala*

Species: *manea*

Subspecies: *schistacea*

Distinguishing characters: Tailed. Under slaty brown with white tinge. Discal band white edged on both sides. In *Rapala,* underhind the discal band is broken at several points. Tornus lobed with tail. Underhind with orange crowned black rectangular spot. Antennal club orange tipped. The underhind end cell bar fainter than the underfore bar. Underhind with two distinct submarginal bands white edged on both sides. Upper blackish brown with ferruginous and bluish shots and white crowned reddish spot at tornus. Body with black and white stripes. Wing span: 28-32mm

Global Distribution: India, Nepal, Bhutan, Myanmer.

Recorded Localities and season: Jakai Biodiversity Park, Dibrugarh, Post Monsoon (August,2023)

Land use and land cover pattern: Forest clearings.

Forage plant: Not recorded

10. Copper flash, *Rapala pheretima petosiris* Hewitson

Sub family: Theclinae

Genus: *Rapala*

Species: *pheretima*

Distinguishing characters: Hind wing with white tipped tails. In *Rapala,* underfore, discal bar, cell end bar and discocellular vein straight. In *pheretima,* above copper reddish. Under is light brown with copper tinge and white edged dark brown marks. The white margin of bar end cell obscure, The discal line underhind catenuated, broken at V3. Spot at tornus prominent, orange obscure, metallic scales at space 1c and 2. Wing span:38-42 mm.

Global Distribution: India, Burma

Recorded Localities and season: Jakai Biodiversity Park. Dibrugarh; Premonsoon (May,2024)

Land use and land cover pattern: Forest clearings, gardens

Forage plant: Not recorded

11. Common Onyx, *Horaga onyx onyx* Moore

Sub family: Theclinae

Genus: *Horaga*

Species: *onyx*

Sub- species: *onyx*

Distinguishing characters: Tailed. Hindwings with three pairs of white tipped tails. In *Horaga,* hind faintly angled at v3 and produced. In *onyx,* upperfore black with dark blue basal patch lined by white discal patch which extends to v1b. Upperhind with blue sheen extending from base to margin bordered by a black apical margin. Thorax with blue scales. Under greenish brown with white patch across both wings. The hind tornus with blue edged black spot. Antennal club yellow tipped. Protected under

Schedule II of IWPA,1972. Wing span:27-31

Global Distribution: India

Recorded Localities and season: Panbari, Kaziranga; Pre Monsoon

Land use and land cover pattern: forest tracks and edge

Forage plant: Not recorded

12. Indian Oak Blue, *Arhopala atrax* Hewitson

Sub family: Theclinae

Genus: *Arhopala*

Species: *atrax*

Distinguishing characters: Blue with black broad apex and termen border. Thorax with bluish scale and hind dorsum dark brown till v1.

Upper hind border black narrow and broadens towards apex. In *Arhopala,* hind with a lobe, and a tooth at v2 . Vein 6 and 7 separate. It was kept under *amblypodia* group in Evans 1932. Antenna with yellow tipped gradual club. The apex of hind is cut off. In *atrax,* upper hind blue only upto cell. Upper purple blue is toothed at end cell.

Global Distribution: India

Recorded Localities and season: Guwahati; Monsoon

Land use and land cover pattern: forest tracks, gardens

Forage plant: Not recorded

13. Common Imperial, *Cheritra freja* Fabricius

Sub family: Theclinae

Genus: *Cheritra*

Species: *freja*

Distinguishing characters: Hind wing with two tails. In *Cheritra,* the tail in V2 is longer than in v1b. Vein 5 is in mid of 4 and 6. Upper brown and under paler brown.In *freja,* underfore with a pale band cell-end. Underhind with narrow post discal dark brown wavy band. In hind a prominent black spot with metallic scales in tornus. Wing span:38-42mm.

Global Distribution: India, Bhutan, Myanmar

Recorded Localities and season: Dibrugarh; Premonsoon.

Land use and land cover pattern: Forest clearings, gardens

Forage plant: Lauraceae

14. Assam Flash, *Rapala tara* De Nicéville

Sub family: Theclinae

Genus: *Rapala*

Species: *tara*

Distinguishing characters: Hind wing with two tails. In *Rapala,* Under end cell bar white edged on both wings followed by a discal band. Hind with orange crowned black spot in tornus. In *tara* upper dark blue . Under, brown with darker brown bands. Discal band underhind broken. Discal band on both wings white edged. Not rare. Observed in large number moving close to ground. Wing span:33-41 mm.

Global Distribution: India, Bhutan, Myanmar, Nepal, Bangladesh

Recorded Localities and season: Jakai Biodiversity Park. Dibrugarh; Premonsoon (May)

Land use and land cover pattern: Forest clearings, gardens

Forage plant: Asteraceae

Subfamily 4 : Lycaeninae (Sapphires)

15. Himalayan Naga Sapphire, *Heliophorus kohimensis* Tytler (F)

Sub family: Lycaeninae

Genus: *Heliophorus*

Species: *kohimensis*

Distinguishing characters: Upper dark. Hind wing tailed at vein 2. Under bright yellow with red marginal band. In *Heliophorus*, the fore v12 is swollen at base and ends after cell-end. The red colour of margin of UNH diffused. UNH end cell indistinct. Underhind white marks anterior to red margin and black marks between margin and base. Under hind a darkbrown spot in cell. In female *kohimensis*, the tornus of UNF bears black spot with broken white marginal boundary. Upper Wing span: 32mm

Global Distribution: India, Bhutan

Recorded Localities and season: Bokakhat, Dibrugarh, Sivasagar; Pre Monsoon

Land use and land cover pattern: Forest tracks and edge, open garden.

Forage plant: Weeds

16. Himalayan Naga Sapphire, *Heliophorus kohimensis* Tytler (M)

Sub family: Lycaeninae

Genus: *Heliophorus*

Species: *kohimensis*

Distinguishing characters: Upper dark with forewing having a shiny purple broad discal patch . Upper hind with lighter purplish discal patch and orange curved marks in space 1-3. Upper, both wings with white marginal bands between veins. In male *kohimensis*, the UPH orange marks reduced and the fore apex rounded. Under both bright yellow. UN with red marginal band. UNF with a black tornal patch uniform white margins. Hind wing with tail and tooth, Underhind, a black spots in cell. Antennal flagellum with white and black patch and a black club. The thorax and legs mottled white, the legs have blackish spots, Wing span: 32mm

Global Distribution: India, Nepal, Bhutan, SE Asia

Recorded Localities and season: Kaziranga; Dibrugarh (University), Sivasagar; Pre Monsoon, Monsoon

Land use and land cover pattern: forest tracks and edge.

Forage plant: Weeds

17. Himalayan Purple Sapphire, *Heliophorus epicles* Godart

Sub family: Lycaeninae

Genus: *Heliophorus*

Species: *epicles*

Distinguishing characters: Upper reddish brown. Hind wing with white tipped tail. UN bright yellow with red margin, no discal bar except some spots between marginal border and base. UNF black tornal spot with white border. In *epicles*, UPF with a broad orange area beyond the cell. UPH with bright red margin. UPF and UPH with marginal white marks. Triangular shaped dark brown marks in the margin of upper hind wings. Wing span: 40mm

Distribution: North East India , Bhutan

Recorded Localities and season: Kaziranga; Dibrugarh (University campus, Dibrugarh; Pre Monsoon

Land use and land cover pattern: forest tracks and edge, open garden.

Forage plant: Weeds

Subfamily 5: Polyommatinae (Line Blues, Hedge Blues etc.)

18. Common Quaker, *Neopithecops zalmora* Butler

Sub family: Polyommatinae

Genus: *Neopithecops*

Species: *zalmora*

Distinguishing characters: Above brown, hind upper margin with dark brown marks between veins. Body brown. In *Neopithecops,* upper v11 and 12 are found separate. In *zalmora,* under white with brown marginal marks between veins. Underhind with prominent black costal spot towards apex and a smaller dark spot in space 1. Wing span: 20-30mm

Distribution: North East India , Bhutan

Recorded Localities and season: Kaziranga; Pre Monsoon

Land use and land cover pattern: forest tracks and edge, open garden.

Forage plant: Weeds

19. Himalayan Common Hedge Blue, *Acytolepis puspa gisca* Fruhstorfer

Sub family: Polyommatinae

Genus: *Acytolepis*

Species: *puspa*

Sub- species: *gisca*

Distinguishing characters: Tailless blue. Upper blue and under pale brown. In *Acytolepis*, underfore bears no costal black dots. The underhind with black distinct spot near costal margin and towards apex. In *puspa*, under fore with distinct basal line in cell-end, followed by a row of elongated black marks oriented in different directions making an irregular discal band. The underhind with a basal black mark like the haltere of a dipteran fly. Underhind prominent white edged basal spot in v. space 7, discal spot in v.space 4 is oblique and elongate. Wingspan:28-35

Global Distribution: India, Nepal, Bhutan, Myanmar

Recorded Localities and season: Kaziranga , Dibrugarh

Land use and land cover pattern: Forest tracks, gardens , grass patches

Forage plant: flowers of grasses

20. Lesser Grass Blue, *Zizina otis* Fabricius

Sub family: Polyommatinae

Genus: *Zizina*

Species: *Otis*

Distinguishing characters: Tailless. Upper darker. Under greyish brown with white edged black marks and spots. Underfore no costal spot. Under hind with two separately placed costal spots. Underhind cell with two spots and a thin end cell bar. Underhind the spot in 4 is oblique and the one in 6 is out of line. Three basal black spots underhind. The species was observed lying over a mollusc with a broken shell. Wing span: 23-24mm

Global Distribution: India

Recorded Localities and season: Kaziranga; Monsoon

Land use and land cover pattern: forest tracks and edge.

Forage plant: The species was found on dead broken mollusc

21. Dark Grass Blue, *Zizeeria karsandra* Moore

Sub family: Polyommatinae

Genus: *Zizeeria*

Species: *karsandra*

Distinguishing characters: Tailless. Upper shiny purplish blue and black apex. Under, greyish brown with white edged well defined dark marks or rounded spots. Upperfore black apex continued to termen. In *Zizeeria*, underfore wing with double line cell-end, a costal spot and a spot in cell. Underfore discal band spot in space 2 is directed inwards. Underhind spots in 3,4,5 are contiguous. Two costal spots underhind. Wing span: 20mm

Global Distribution: India

Recorded Localities and season: Guwahati, Dibrugarh ; Monsoon, postmonsoon

Land use and land cover pattern: Urban garden, park, herbs and shrubs

Forage plant: *Helianthus* spp., flowers of Asteraceae

22. Pale Grass Blue, *Zizeeria maha* Kollar (M)

Sub family: Polyommatinae

Genus: *Zizeeria*

Species: *maha*

Distinguishing characters: Tailless. Upper milky blue with palr brown border along termen on fore and narrow border on hind. IIn *Zizeeria*, underfore wing with double line cell-end, a costal spot and a spot in cell. UNH a spot in 1a below spot mid cell. In *maha*, underfore with a spot in cell and the spot in vs7 lies between cell end bar and discal band spot 6. Underhind marks obsolete. In this subspecies underhind is with two costal spots, spot at 6 aligned and submarginal spots pale. Wing span: 26-30mm

Global Distribution: India

Recorded Localities and season: Dibrugarh; Winter (December)

Land use and land cover pattern: Urban garden, park, low cover grasses

Forage plant: Flowers of Asteraceae

23. Pale Grass Blue, *Zizeeria maha* Kollar (F)

Sub family: Polyommatinae

Genus: *Zizeeria*

Species: *maha*

Distinguishing characters: Tailless. Upper shiny brown with dark brown vein. Under, greyish brown with white edged dark marks. In *Zizeeria*, underfore wing with double line cell-end and a spot in cell. UNH a spot in 1a below spot mid cell. In fore, origin v2 is nearer base than origin v11. Underfore discal band spot in space 4 is directed outwards. In *maha*, below markings prominent. Underfore with a spot in cell and the spot in vs7 lies between cell end bar and discal band spot 6. Two costal spots underhind. Basks keeping wings open. In this subspecies underhind is with two costal spots, spot at 6 aligned and submarginal spots pale. Wing span: 26-30 mm

Global Distribution: India

Recorded Localities and season: Dibrugarh ; postmonsoon (August)

Land use and land cover pattern: Urban garden , park, low cover grasses

Forage plant: Flowers of Asteraceae

24. Himalayan Pale Grass Blue, *Pseudozizeeria maha maha* Kollar

Sub family: Polyommatinae

Genus: *Pseudozizeeria*

Species: *maha*

Subspecies: *maha*

Distinguishing characters: Tailless. Under pale greyish brown. In *Pseudozizeeria*, underhind, spot in 6 is out of line. In *maha maha*, under, prominent white edged rounded black marks. Under cell with black spot and a bar in cell-end. Underfore a costal spot between spot in space 7 and mark cell-end. Underhind in base of v. space1b, a very small black spot is present. Margins with uniform cilia. Under, submarginal spots distinct.

Wing span: 26-30 mm

Global Distribution: India

Recorded Localities and season: Kaziranga ; postmonsoon

Land use and land cover pattern: Urban garden , park, herbs and shrubs, forest tracks

Forage plant: Flowers of herbs

25. Plain Hedge Blue, *Celastrina lavendularis* Moore

Sub family: Polyommatinae

Genus: *Celastrina*

Species: *lavendularis*

Distinguishing characters: Tailless. Upper brown with blue suffusion and without any white patch. Under greyish brown. Underhind with distinct submarginal spots, the spot in1a is in mid dorsum. Under, spots are broader and variable. Wing span:28-34mm

Global Distribution: India

Recorded Localities and season: Jakai Biodiversity Park, Dibrugarh,

Spring (May)

Land use and land cover pattern: Urban garden, park, herbs and shrubs, forest tracks

Forage plant: Flowers of herbs

26. Tailless Line Blue, *Prosotas dubiosa indica* Evans

Sub family: Polyommatinae

Genus: *Prosotas* (= *Nacaduba*)

Species: *dubiosa*

Sub- species: *indica*

Distinguishing characters: Under dark brown with orange crowned black spot in space1, 2. In *Prosotas,* tail absent . Under, pairs of white edged lines broken into fragments In *dubiosa,* base of wings bluish, bands slightly darker and very irregular. Under wing margins with pale brown crowned dark spots. Marginal cilia uniformly brown with bluish tinge. Antennal club white tipped. It is protected under Schedule II of IWPA, 1972 . Wing span:22-26mm

Global Distribution: India

Recorded Localities and season: Kaziranga; Monsoon

Land use and land cover pattern: forest tracks and edge. Attracted to human sweat.

Forage plant: Not recorded

27. Brown Line Blue, *Prosotus lutea* Martin

Sub family: Polyommatinae

Genus: *Prosotus*

Species: *lutea*

Distinguishing characters: Tailless. Under brown with black bordered dark patches in line or fragments. Under with marginal black border and submarginal white spots bordered marginally by black lines and apically by dark brown lines. In *lutea,* under paired dark brown lines. Under hind one submarginal white spot is covered by distinct black spot in space2. The underhind apical black mark large. Under discal and basal band not in continuous line. Underfore double bar in cell-end. Wing span: 24-28

Global Distribution: India

Recorded Localities and season: Kaziranga, Jakai Biodiversity Park, Monsoon, Pre monsoon (May)

Land use and land cover pattern: forest tracks and edge.

28. Bengal Common Ciliate Line Blue, *Anthene emolus emolus* Godart

Sub family: Polyommatinae

Genus: *Anthene* (= *Lycaenesthes*)

Species: *emolus*

Subspecies: *emolus*

Distinguishing characters: Tailless. Under dull brown with pale lines. In *Anthene,* cilia in hind in v. space1,2,3 longer. Under paired white lines forming bands. In *emolus emolus,* under discal band more or less continuous. Underhind one submarginal prominent orange crowned black spot in v.space2 and one prominent basal black spot in dorsum. Cilia in fore uniform. Antennal club brown tipped. Wing span: small

Global Distribution: India

Recorded Localities and season: Kaziranga; Monsoon

Land use and land cover pattern: forest tracks and edge, found attracted to human sweat.

Forage plant: Not recorded

29. Oriental Common Cerulean, *Jamides celeno celeno* Cramer

Sub family: Polyommatinae

Genus: *Jamides*

Species: *celeno*

Sub- species: *celeno*

Distinguishing characters: Tailed. Under greyish brown with submarginal bluish white lines. In *Jamides*, v11 and 12 separate with a small connection. Under fore white line endcell is straight. In *celeno*, under fore with three costal dashes above lines end cell. Underhind tornus with orange crowned black spots till v. space 2 . The pair of white lines above tornal spot are sharply 'v' shaped. Submarginal pair of lines underfore catenuated but continuous. Wing span: 27-40 mm.

Global Distribution: India

Recorded Localities and season: Kaziranga; Monsoon

Land use and land cover pattern: forest tracks and edge

Forage plant: *Gossypium*

30. Plains Cupid, *Chilades pandava* Horsefield

Sub family: Polyommatinae

Genus: *Chilades*

Species: *pandava*

Distinguishing characters: Tailed. UN light silver brown with white bordered discal bar. UNH large orange crowned black spot at tornus and two white bordered black costal spots. Upper blue with black outer margin. Veins black. In *pandava,* four black spots at UNH base. Wing span:25-35 mm.

Global Distribution: India, Pakistan Sri Lanka, Nepal, Bhutan

Recorded Localities and season: Guwahati

Land use and land cover pattern: Open garden

Forage plant: Poaceae

31. Common Pierrot, *Castalius rosimon rosimon* Fabricius

Sub family: Polyommatinae

Form II

Genus: *Castalius*

Species: *rosimon*

Sub species*: rosimon*

Distinguishing characters: Tailed. Upper base of wings bluish. Under fore with basal two marks arranged in the form of inverted 'V'. In *Castalius*, v 11 ends beyond cell-end in forewing. In *rosimon rosimon*, upper white with black marks. Underfore, a long basal streak continued with cell-end mark , a cell mark extended beyond cell boundary and a basal mark in v. space 1b,all well separated. Underhind with central region without marks. Upper fore with basal blue region and black broad margin. Rarely they bask with wings open. Wing span :20mm

Two forms are noticed. In form II, UNF discal marks in v.space 1b,2,3,4 ,5 all well separated, mark in space 3 and 5 are coalesced with submarginal band.

Global Distribution: India, Pakistan, Nepal, Bhutan, Bangladesh

Recorded Localities and season: Guwahati, Dibrugarh ; Monsoon

Land use and land cover pattern: Urban garden, basks above.

Forage plant: *Alternanthera* spp., *ocimum* spp

32. Little Tiger Pierrot, *Tarucus balkanica* Freyer

Sub family: Polyommatinae

Genus: *Tarucus*

Species: *balkanica*

Distinguishing characters: Upper purplish blue basally, brownish toward margin with marks and dark brown veins. Under fore with two marks making an inverted 'V' shape. *Tarucus* is blue pierrot with under marks in the form of streaks except the marginal marks. Under both wings have basal streaks. In *balkanica*, underfore spot in midcell is equidistant from basal streak and streak cell-end. Underfore with two uniformly arranged rows of submarginal spots. Underhind, the marginal marks with metallic scales and of different shapes. Moves fast over grasses in bright sunlight, sits with folded wings. Wing span: 20 mm

Global Distribution: Assam, Sikkim

Recorded Localities and season: Dibrugarh (University), November

Land use and land cover pattern: Understory and open herbs

Forage plant: *Alternanthera* spp.

33. Zebra Blue, *Syntarucus plinius* Fabricius

Sub family: Polyommatinae

Genus: *Syntarucus*

Species: *plinius*

Distinguishing characters: White with alternate dark brown and light

marks. Under both wings have basal streaks. Underfore cell mark is connected with basal streak. Underhind alternate irregulat broad and narrow streaks broken into spots. Wing span:24mm

Global Distribution: India

Recorded Localities and season: Dibrugarh; Winter (December)

Land use and land cover pattern: Urban garden, park, low cover grasses

Forage plant: Flowers of grasses

34. Dark Pierrot, *Tarucus ananda* de Nicéville

Sub family: Polyommatinae

Genus: *Tarucus*

Species: *ananda*

Distinguishing characters: Tailed. Under bluish white with dark brown marks. *Tarucus* is blue pierrot with basal streak in both wings. In *ananda*, underhind discal spot at v.space 5 is coalesced with post-discal band. In difference to Evans (1932) however, undefore discal spot at v.space 5 is coalesced with spot 4 and 6 rather than with post-discal band. Abdomen with black and white stripes. The species *ananda* is Legally protected under schedule IV of Indian wildlife protection Act,1972. Wing span: 22-28mm

Global Distribution: India

Recorded Localities and season: Kaziranga; Pre Monsoon

Land use and land cover pattern: forest tracks and edge

Forage plant: *Laportia* spp.

35. Indo Chinese Elbowed Pierrot, *Caleta elna noliteia,* Fruhstofer

Sub family: Polyommatinae

Genus: *Caleta*

Species: *elna*

Subspecies*: noliteia*

Distinguishing characters: Upper black with white central discal band. The under wings have thin black marginal band with white spots. In *Caleta,* upperhind discal band broad and extends from costa to dorsum. In upperfore, the white discal band is narrower and extend from dorsum to v.space 4, obsolete in 5. In *elna noliteia* , underfore with an angled basal broad band from costa to dorsum meeting the black basal band of underhind. The underfore discal band is divided into three parts due to shifting of middle mark outward to coalesce with marginal border. The under hind has thick discal black band. The base and space between marginal band and discal band is pale-yellow. The thorax, legs and abdomen is black with white marks. Found getting attracted to human sweat. Wing span: 30mm

Global Distribution: India

Recorded Localities and season: Jakai Biodiversity Park, Dibrugarh, Gibbon wild life sanctuary. Stream side, grasses. Premonsoon, Monsoon

Land use and land cover pattern: forest clearings, parks, gardens

Forage plant: *not recorded*

36. Indian Red Pierrot, *Talicada nyseus nyseus* Guérin-Méneville

Sub family: Polyommatinae

Genus: *Talicada*

Species: *nyseus*

Sub-species: *nyseus*

Distinguishing characters: Tail in hind v2 . Above black with broad orange area in hind. In *Talicada,* margin is chequered and under white with large discal black spots. In *nyseus,* marginal border underfore black with white spots and underhind orange till v.space 5 with white spots. Underhind dark prominent spot in cell and a mark in cell-end. Underfore, a basal dark streak. In *nyseus nyseus,* discal band underhind reduced or absent. Wing span:37mm.

Global Distribution: India, Bhutan, Myanmar

Recorded Localities and season: Guwahati, Kokrajhar, Dibrugarh; Monsoon

Land use and land cover pattern: Forest tracks, gardens

Forage plant: *Helianthus, Catharanthus*

37. Pea Blue, *Lampides boeticus* Linnaeus

Sub family: Polyommatinae

Genus: *Lampides*

Species: *boeticus*

Distinguishing characters: Upper brown with blue flash at base of wings. In *Lampides,* hind tail long and at V2. Under catenuated white edged dark brown bands. Hind on both sides with two black spot in tornus, The latter

underhind are orange crowned. In *boeticus*, under fore cell with five dark brown bars and no bar at the level of cell. Wing span: 24-36mm.

Global Distribution: India, Bhutan, Myanmar, Ceylon, Andaman Nicober.

Recorded Localities and season: Dehing Patkai; Spring (April)

Land use and land cover pattern: Forest tracks, gardens

Family: Riodinindae

1. **Punchinello, *Zemeros flegyas* Cramer**

    Subfamily: Nemeobiinae (= Riodininae)

    Genus: *Zemeros*

    Species: *flegyas*

Distinguishing Characters: Body brown with black scales. Upper brown with black elongated marks bordered by white. Upper fore and hind with black margin and white spots. In *flagius*, upper cells with one white mark. Wingspan: 40 mm. Flight rapid.

Global Distribution: India, Sikkim, Parts of Oriental region

Recorded Localities: Jakai Biodiversity Park, Kaziranga, Kokrajhar, Udalguri.

Land use land cover pattern: Mostly open places near forest, close to ground

Forage plant: *Knoxia* spp.

References

1. Condamine, F. L., Allio,R,. Reboud, E. L., Dupuis, J. R ., Toussaint, E. F. A., Mazet, N., et al. (2023) A comprehensive phylogeny and revised taxonomy illuminate the origin and diversification of the global radiation of *Papilio* (Lepidoptera: Papilionidae). Molecular Phylogenetics and Evolution, 107758.
2. Evans, W. H. (1932). The identification of Indian butterflies. The Bombay Natural History Society. India.
3. Evans, W. H. (1949). A catalogue of Hesperidae in Europe, Asia and Australia in the British Museum (Natural History). Trustee of the British Museum. London.
4. Haribal, M. (1992). The butterflies of Sikkim Himalaya and their natural history. Sikkim Nature Conservation Foundation. Sikkim.
5. Joshi, J., and Kunte,K. (2022) Polytypy and systematics: diversification of *Papilio* swallowtail butterflies in the biogeographically complex Indo-Australian Region. bioRxiv, 2022.03.23.485569.
6. Khemikar, I. (2008). The book of Indian butterflies. Bombay Natural History Society. India.
7. Kunte, K. (2007). India-a lifescape butterflies of peninsular India (M. Gadgil & E. O. Wilson, Eds.). Universities Press.
8. Kunte, K., Sondhi, S., and Roy, P. (Chief Editors). (n.d.). Butterflies of India, v. 3.06. Published by the Indian Foundation for Butterflies. Retrieved from https://www.ifoundbutterflies.org/

7\. Labandeira, C. C., Yang, Q., Santiago-Blay, J. A., Hotton, C. L., Monteiro, A., Wang, Y.-J., Goreva, Y. S., Shih, C., Siljeström, S., Rose, T. R., Dilcher, D. L., & Ren, D. (2016). The evolutionary convergence of mid-Mesozoic lacewings and Cenozoic butterflies. Proceedings of the Royal Society B: Biological Sciences, 283(1824). https://doi.org/10.1098/rspb.2015.2893

8\. Nandi, S. (2021). *Charaxes marmax* Westwood, 1847 - Yellow Rajah. In K. Kunte, S. Sondhi, & P. Roy (Chief Editors), Butterflies of India, v. 3.21. Indian Foundation for Butterflies. Retrieved from http://www.ifoundbutterflies.org/sp/914/Charaxes-marmax

9. Natural History Museum, London. Global Lapidoptra name index nhm.ac.uk/our-science/data/lepindex/help

10. Palma-Onetto, V., Bergmann, J., & González-Teuber, M. (2023). Mode of action, chemistry and defensive efficacy of the osmeterium in the caterpillar *Battus polydamas archidamas*. Scientific Reports, 13, 6644. https://doi.org/10.1038/s41598-023-33390-x

11. Petschenka, G., and Agrawal, A. A. (2015). Milkweed butterfly resistance to plant toxins is linked to sequestration, not coping with a toxic diet. Proceedings of the Royal Society B, 282, 20151865. http://dx.doi.org/10.1098/rspb.2015.1865

12. Wynter Blyth, M.A. (1957) Butterflies of the Indian Region. Bombay Natural History Society, Bombay, India

CHAPTER IV

Ecology and Ecological significance of Butterflies

Dipsikha Bora, Anisha Phukan and Jyoti Agrahari

The intricate relationship of organisms with the environment consisting of the biotic and abiotic components constitutes the fundamentals of science of ecology. The ecological interactions butterflies enter into involve consumption of host plants including seasonal flowers, crops, forest trees, pollination, nutrient cycling, food chain and other ecosystem processes. While admiration for butterflies mostly roots in their beautiful physical design and love for nature, deeper understanding explains the worth of its presence in nature. The biomass, that the magnificent butterflies contribute is also impressive. The state of Assam belongs to the Eastern Himalaya Biodiversity region and endowed with diverse ecological conditions like grasslands, tropical wet evergreen, semi evergreen, moist deciduous, subtropical hill forests and marshy lands with spectacular number of flora which makes the region a paradise for butterflies.

Habits and habitat

Most butterflies are diurnal, meaning they are active during the daytime. They rely on sunlight and warmth for flight and feeding. The most important factor influencing butterfly habitat is its food source. Adult butterflies primarily feed on nectar from flowers. Host plants of butterflies are species-specific. Adults sip nectar from the flowers of specific plants. The caterpillars feed on their specific hostplants and the female adult lays eggs on the plant where their larvae can feed and survive.

Mud-puddling is one of the intriguing ways butterflies interact with their environment, showcasing their adaptability and ecological significance. Some butterflies, gather on damp soil, mud, or mineral-rich substrates to extract essential minerals and nutrients vital for their health. This behavior is commonly observed in male butterflies, although some females may also participate[1]. Examples of butterflies observed being engaged in mud puddling include the emigrants, common tiger, swallowtails, painted lady, and sulfur butterflies.

Butterflies often perch and bask in the sun with their wings open. They are poikilothermic and hence their body temperature change along with environmental temperature. At cool temperature, basking helps them raise their body temperature which also aids in flight performance. For example,

Pieris requires thoracic temperature of 29° and 40°C to take off and fly and they achieve these elevated temperatures by basking [2]. During basking, butterflies may be seen to change body plane angle facilitating heating of different parts. The wings have scales with different degree of pigment deposition which cause differential absorption of visible and infrared wavelength. Similarly different parts of the butterfly wings also have different emissibility which prevents overheating of wing parts like scent patch, scent pad and veins during basking. Thus, optimum temperature is maintained in different body parts for day time activities [3].

Butterflies engage in courtship rituals, involving specific behaviors and displays to attract the mates. The author has observed in the case of common mormorn (*Papilio polytes*) making flight in the air initially by making circular movements around each other at different heights followed by wing fluttering during body orientation before courtship. Some butterfly species undertake long-distance migrations to escape adverse conditions or find suitable breeding and overwintering grounds.

Some common habitats harbouring butterflies include open grasslands, meadows, marshes, swamps, riverbanks, woodlands, forest edges, mountains and hillsides, coastal areas, agricultural fields, gardens and parks . Variation in seasonal occurrence of butterflies is common. In Assam, during the winter months of December-January the population of adult butterfly becomes significantly less in open areas, parks and gardens. In all other months, based on availability of nectaring flowers their population fluctuate. In urban areas the most common are the albatross, grass yellow, psyche and cabbage white which are found almost in all seasons.

Feeding interactions

Butterfly adults mostly rely on nectar of flowers while some feed on cadavers, animal droppings , rotten foliage and fruits, minerals in wet soil and even human sweat. For nectaring, many butterflies prefer plants with composite flower probably for minimising energy loss. The larvae feed mostly on the leaves and few like some Lycaenids feed on flower buds or petals. Some are oligophagous, feeding plants of the same family and the others are polyphagous, feeding on plants of different families. Many Hesperids prefer only plants of Poaceae family. Danaids mostly prefer the milkweeds, *Asclepius* (Apocynaceae). Polyphagy among butterflies is more common. The caterpillar feeding cause changes in the chemistry of the plants and sometimes influence community interactions. While the caterpillars are herbivores, they themselves serve as prey for primary and secondary carnivores like reptiles, birds, mammals and also for parasitoids. The Eastern Tailed-Blue butterfly (*Cupido comyntas*) is commonly hunted by birds like sparrows, warblers, and flycatchers for their high-protein content [4]. Orb-weaving spiders like *Argiopes* which are very common in urban gardens of Assam may be easily seen to catch butterflies in their intricate webs of silken threads. Certain parasitoid wasps, such as the Braconidae and Ichneumonidae families, lay their eggs on or in butterfly caterpillars. The wasp larvae then feed on the caterpillar, eventually killing it [5]. However, some butterfly caterpillars like those of Danaids when feed on milkweed species having more cardeneloids contain more toxins in their body. The predatory birds avoid them by way of a learning process.

Interspecific interactions

Mutualism: Mutualistic association between two species evolve when two free-feeding species evolve character or behaviour which are of great selective advantage to both. Mutualism develops between taxonomically divergent species belonging to different trophic levels using different resources. Lycaenid butterflies exhibit a very well known phenomenon of mutualism with ants. The caterpillar maintaining such relationship bears sugary and dew patch glands through which they secrete nectar. The ants avail a significant amount of their energy through nectar while protecting the tiny caterpillars of myrmicophilous Lycaenid butterflies. Recently, a Riodinid butterfly, equipped with perforated cupola organs and tentacle nectary organs exhibiting mutualistic behaviour with ant has been found to reside inside ant colony [6].

Butterflies also maintain a mutualistic association with flowers. The length of proboscis determine the flower they visit for nectaring to obtain necessary nutrients. Certain plants have developed unique flower structures that can only be accessed by a particular butterfly species, ensuring future visit of the butterfly. The long proboscis of many Hesperids allow them to nectar on flowers having long corolla. Such specialized relationships contribute to the maintenance of plant diversity, promote co-evolution between butterflies and their host plants and provide stability to the biotic communities. From the flower nectar, the butterflies obtain reliable food resource. But while nectaring they transfer pollen between flowers. Butterflies do not have specialized structures for collecting and storing pollen. Their bodies lack the branched hairs, called scopae, that bees use to carry pollen. Instead, butterflies rely on the electrostatic forces that cause pollen grains to stick to their bodies. This method of pollen transfer, though less efficient enables butterflies to contribute significantly to plant reproduction. Research concerning functional role of pollination service of butterflies reveal plant species as a feeding source to be the limiting factor [7].

Mutualism is essential in conservation of butterfly diversity.

Maintenance of natural pollinating system and vegetation for preserving ant habitat are critical for survival of mutualistic relations. Reduction of suitable site due to urbanization and unplanned deforestation may lead to loss of butterfly fauna.

Commensalism: Commensalism is a symbiotic relationship between two species in which one avails the benefit without harming the other. The one drawing the benefit is called the commensal while the one not affected is called the host. The caterpillars of monarch butterflies (Danaids) feeding on the milkweeds store the toxic cardiac glycosides in their bodies and avoid predation. Thus, the monarch butterflies act as commensals.

Antipredator defensive adaptation through mimicry: Predation is the interspecific interaction in which the prey is consumed by the predator. The naturalist HW Bates in 1862 first described the Batesian mimicry. A palatable prey uses deceptive means as a best defensive strategy to avoid detection and predation. The palatable prey resemble the unpalatable prey and the resemblance evolved because of its advantage to the resembling group . For example, the early instars of *Papilio* species resemble bird droppings. In such deceptive strategy, however, efficiency may decrease with increase in density of the mimic and degree of palatability depending on learning experience of the predator. A predator having a learning ability is likely to avoid the butterfly that resembles the unpalatable one. But if a rare new phenotype resembling a different model species arises, the fitness will increase. Examples of mimicry among butterflies are exhibited by Great egg fly males resembling the Danaid Eggfly males and Great egg fly female resembling a female *Euploea* (crow).

The unpalatable prey on the other hand avoid attack through aposematism. The phenomenon is used to advertise unpalatability due to the presence of toxic substance in the body through conspicuous characters such as warning colouration. The predator detect, learn, recognize and avoid. Such aposematism is exhibited by Danaid females which are polymorphic with conspicuous warning colouration.

In the case when the prey carries toxic substance, the predator has to taste them to understand that the food resource is toxic and need avoidance. This concept was formalized by Fritz Muller in 1879 and is called as Mullerian mimicry. He started similarly with Bates that "Mimicry arose as a consequence of selection imposed by predation" [8]. When two warning species are subjected to the same group of predators they may share the same warning signal. Species sharing the same warning is called co-mimetic, interact mutualistically and form mimetic ring that express convergence of warning due to local predation pressure. But the intensity of warning or convergence depends on the exposure of prey to the predator class. Segregation of predators in microhabitats differentially influences the adaptive warning of the prey. Mimicry also has other ecological implications. Altitudinal segregation of mimetic forms are also in report. With achange in altitude or climatic niche, mimetic rings (Large collection of similar looking species) may get segregated along with change in microhabitats.

Mullerian mimetic species have long coevolutionary history. The mimicry is comparatively stable and conditioned by other traits used as mating cues, the genetic architecture of colour pattern elements and the ecological niche of the species. In *Heliconius* butterflies, homologous chromosomal regions control diversity in mimicry. A restricted set of genomic regions are responsible for warning red colour, melanism in proximal wing part and in median part of fore and hind wing. The genetic

Great Eggfly *(Hypolimnas bolina)* (Male)

Danaid Eggfly *Hypolimnas misippus* (Male)

Stripped tiger*(Danaus genutia)*

Plain tiger*(Danaus chrysippus)*

architecture though is ancestral and conserved, the mimetic characters between lineages have evolved through convergent evolution [9].

Butterflies also mimic background colouration and avoid detection. Several nymphalids exhibiting seasonal forms provide examples of such mimicry.

Ecosystem services: Apart from pollination and providing food for others, butterflies also play a vital role in nutrient cycling through herbivory.

Caterpillars consume vast amounts of plant material, which helps to recycle nutrients back into the soil through felling of leaves, their faeces and dead bodies, supporting plant growth and overall ecosystem health.

Migration

Butterfly migration is a natural phenomenon where butterflies travel long or short distances. Through their migrating behaviour butterflies influence their population density in a specific area and also ecosystem services. Migration is typically driven by seasonal changes and the need to find suitable conditions for food, survival and reproduction. Changes in day length, temperature, and the scarcity of specific host plants for caterpillars play influential role [10]. Not all butterfly species migrate, but some well-known migratory butterflies include the Monarch butterfly (*Danaus plexippus*), the Painted Lady butterfly (*Vanessa cardui*), and the Red Admiral butterfly (*Vanessa atalanta*). Four Danaine species, namely, Dakhan Dark Blue Tiger (*Tirumala septentrionis dravidarum*), Oriental Blue Tiger*(Tirumala limniace exoticus),* Double-branded Black Crow *(Euploea sylvester coreta*), and Indian Common Crow *(Euploea core core)* form swarms that migrate across the plains of Southern India to the southern part of Western Ghats between mid-October and early December every year to avoid heavy monsoon[11]. Crimson rose butterflies (*Pachliopta hector*) migrate from southern India across the sea to Sri Lanka in February. Another example is the Monarch butterfly subspecies, *Danaus plexippus plexippus* which migrates across North America each summer and autumn to and from overwintering locations in California and Central Mexico[12]. The journey can cover thousands of miles and involves multiple generations of butterflies. In migratory species like Monarchs, individuals that migrate long distances are not the same butterflies that return. Instead, they go through several generations during the migration process. Although it's still a subject of study, butterflies are believed to use a combination of cues for navigation during migration. These cues include the position of the sun, the earth's magnetic field, visual landmarks, and possibly even their sense of smell. Butterflies often follow specific migratory corridors, which are well-established routes between their breeding and overwintering sites. These corridors may vary depending on the species and their geographical distribution. Protecting key habitats along their migration routes is essential for ensuring the survival of migratory butterfly populations. In Assam , butterflies are observed to fly around for distances out and in of the reserved forests. Although no spectacular migratory swarms are recorded so far, there are reports of altitudinal migration and winter migration to Assam from Bhutan-Himalaya. Lycaenids migrate from high elevation to low elevation during winter and vice versa during summer in Barail Hill range, a south western extension of Patkai range [13]. Studies conducted in upper Brahmaputra valley of Assam show that autumn temperature period has extended over the past decade and has reduced winter duration. Therefore, migration of butterflies from cold hilly zones to upper Assam is possible. An increase of autumn temperature period and migration from cooler regions might be the reason for sighting of certain species like *Pieris* even during winter in upper Assam.

Butterfly as indicator of environmental stress

Butterflies have been extensively studied primarily because of their magnificent appearance and obvious presence in different habitats. Hence they have well-defined taxonomy, life history, and biology details for many. They occupy a low trophic level in the ecological pyramid and are dependent for survival on a plant-specific relationships. Their activities and flight periods are dependent on climatic factors like seasonal temperature, wind, rainfall and photoperiod. Therefore, they are vulnerable to extreme changes in environmental factors. Shifts in the ranges of certain species or alterations in their flight periods can be early warning signs of climate-related changes. The range expansion of the European butterfly species, the comma butterfly (*Polygonia c-album*), into northern regions of the UK has been linked to milder winters, which can be considered an indicator of changing climate [14].

Loss of heterogeneous vegetation in degraded forests of Assam has been reported to cause loss of forest butterflies in Assam [15] which can adversely affect ecosystem health. Monitoring the recovery of endangered or threatened butterfly species can provide insights into the effectiveness of conservation efforts and the overall health of their habitats. The successful reintroduction of the Oregon silverspot butterfly (*Speyeria zerenehippolyta*) into its native habitat in the Pacific Northwest serves as an indicator of the effectiveness of conservation efforts in preserving this endangered species [16]. Establishment of biodiversity parks in the buffer zones of the protected areas can prove to be an effective way conserving forest butterflies and monitoring environmental impact as well.

Butterflies can also serve as indicator of environmental pollution. The decline of the Monarch butterfly (*Danaus plexippus*) has been linked to the use of herbicide, glyphosate, which kills milkweed plants, the primary food source for Monarch caterpillars [17]. The loss of butterflies in Assam due to pesticide use is not yet estimated.

References

1. Beck, J., Mühlenberg,E., and Fiedler,K., (1999). Mud-puddling behavior in tropical butterflies: In search of proteins or minerals?, Oecologia, 119 (1)140-148, doi: 10.1007/s004420050770.
2. Kingsolver, J.G., (1985).Thermal ecology of *Pieris* butterflies (Lepidoptera: Pieridae): a new mechanism of behavioral thermoregulation Oecologia, 66(4):540-545. doi: 10.1007/BF00379347
3. Tsai, C.C., Richard, A., Childers, N.N. S., Crystal, R. , Julianne, N.P., Bernard, G.D.,Naomi, E.P. and Nanfang, Yu. (2020). Physical and behavioral adaptations to prevent overheating of the living wings of butterflies . Nature communication, 11, 551. https://doi.org/10.1038/s41467-020-14408-8
4. Robbins, C. S., Bruun, B., and Zim, H. S. (1983). A Guide to Field Identification: Birds of North America. Golden Guides from St. Martin's Press.
5. Gauld, I. D., and Bolton, B. (1988). The Hymenoptera. Oxford University Press.
6. Pérez-Lachaud., G., Rocha, F.H., Pozo, C. et al. (2021). A new ant-butterfly symbiosis in the forest canopy fills an evolutionary gap. Sci Rep 11, 20770. https://doi.org/10.1038/s41598-021-00274-x
7. Cristian, A., Martínez-Adriano., C,. Díaz-Castelazo and Aguirre-Jaimes., A (2018). Flower-mediated plant-butterfly interactions in a heterogeneous tropical coastal ecosystem. Peer J. DOI 10.7717/peerj.5493
8. Müller, F., Ituna, and Thyridia, (1879)A remarkable case of mimicry in butterflies, Proc. Entomol. Soc. Lond. , xx-xxix
9. Marianne, E., and Mathieu, J.,(2015). Mimicry in *Heliconius* and *Ithomiini* butterflies: The profound consequences of an adaptation. BIO Web of Conferences 4, 00008 (2015) DOI: 10.1051/bioconf/20150400008 Published by EDP Sciences.
10. Chowdhury, S., Zaluki, MP.,Amano, T., Poch,T.Z., et al.,(2022). Trends and progress in studying butterfly migration, Integrative Conservation, 1(1), 8-24, doi: 10.1002/inc3.13.
11. Kunte, K.,(2005) Species composition, sex-ratios and movement patterns in danaine butterfly migrations in southern India, J. Bombay Nat. Hist. Soc., 102(3), 280-286.
12. Riley, N.D. (1951) The monarch butterfly, Science, 114 (2951),75-76, doi: 10.1126/science.114.2951.75.
13. Gogoi ,M. (2019) Distribution pattern and species diversity of

butterflies in different habitats along altitudinal gradients in Barail hill range. Ph. D Thesis, Assam University

14. Roy, D.B., Rothery, P., Moss, D., Pollard, E., and Thomas, J. A. (2001). Butterfly numbers and weather: predicting historical trends in abundance and the future effects of climate change. Journal of Animal Ecology, 70(2), 201-217.
15. Saikia, M.K., Kalita , J. and Saikia, P.K. (2009).Ecology and conservation needs of nymphalid butterflies in disturbed tropical forest of Eastern Himalayan biodiversity hotspot, Assam, India International Journal of Biodiversity and Conservation 1(7) ,231-250. http://www.academicjournals.org/ijbc
16. Schultz, C. B., Dlugosch, K. M., & Ehrlich, P. R. (2012). Testing the geographic scale of coevolution in a plant-herbivore interaction: The case of the California butterfly *Euphydryas editha* and its host plant *Pedicularis*. Evolutionary Ecology Research, 14(6), 655-674.
17. Flockhart, D. T., Pichancourt, J. B., Norris, D. R., & Martin, T. G. (2015). Unraveling the annual cycle in a migratory animal: breeding-season habitat loss drives population declines of monarch butterflies. Journal of Animal Ecology, 84(1), 155-165.

CHAPTER V

Study of the Butterflies: Effective Techniques for Collection and Preservation

Anisha Phukan

Butterfly collections done responsibly and ethically can be a fascinating hobby and a valuable way to contribute to teaching, scientific research and nature conservation efforts.

Equipments and materials used in butterfly collection

Butterfly net: A lightweight, fine-meshed net used to catch butterflies without harming them.

Killing jar: A container with a tight-fitting lid filled with a killing agent like ethyl acetate, isopropyl alcohol, or chloroform to euthanize the butterflies quickly and painlessly.

Fruit baits: Some butterflies are attracted to the sugars and aroma released by ripe and decaying fruits. Common fruits used as bait include banana, orange, apple, pineapple, and other soft, juicy fruits. To prepare the bait, selected fruits are cut into small pieces or mashed slightly so that their scents are emanated into the surrounding air. The bait is placed in a shallow container or plate.

Butterflies are more likely to be found in areas well sunlit, with flowering plants nearby, or along the fringe areas of forests. The baits are set up in such areas where frequenting of butterflies are sighted. A net can be used to gently catch the butterflies when they land on the bait.

Low sedatives: Instead of using a killing jar as described above, a more humane method using ethyl acetate-soaked strips may be used that act as a low sedative for such treatment. A small piece of absorbent material such as cotton ball or paper towel is soaked with ethyl acetate, and placed in a tightly sealed container. Alternatively, inside a small jar with a perforated lid may be placed the ethyl acetate-soaked material and put the butterfly inside. The fumes will sedate the butterfly quickly without causing harm as the goal is to sedate the butterfly, not kill it. Once the butterfly becomes immobile, it is to be promptly removed from the container for study. On completion of the process, butterflies need to be released back into their natural habitat after examination.

Spreading board and pins: Boards made of wood or other materials

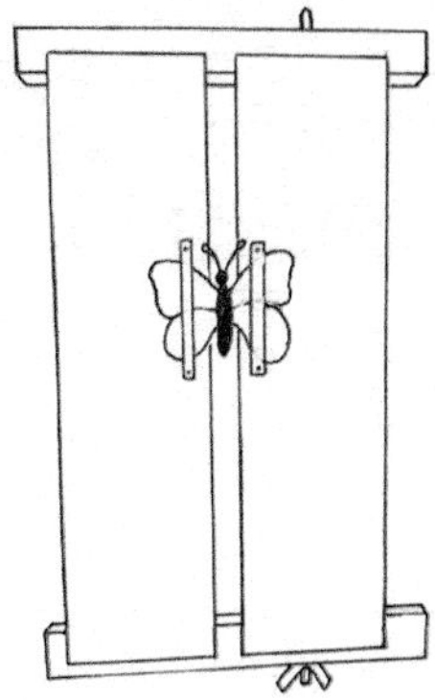

Butterfly spreading tool

Insect killing Bottle

Fine pointed forceps

Butterfly catching net

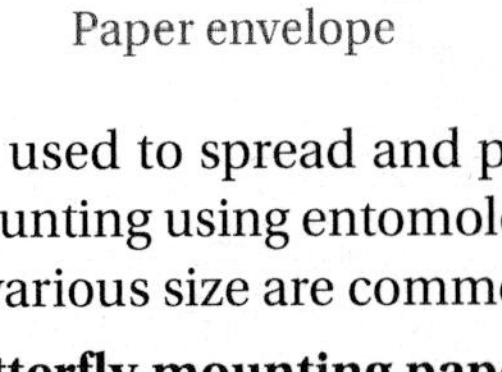

Paper envelope

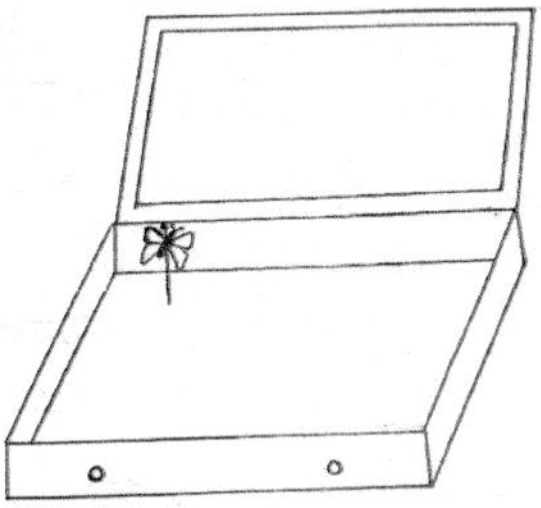

Insect collection box

are used to spread and position the wings of the butterfly for drying and mounting using entomological pins made of rust free materials. Such pins of various size are commercially available .

Butterfly mounting paper or foam: Soft, non-sticky material are used to hold the butterfly in place during drying and mounting.

Entomological forceps: Delicate, fine-tipped tweezers are used to handle the specimens.

Method of Collection

Ethical concerns: While butterfly collection can be a fascinating hobby, it is essential to approach it with ethical considerations in mind. It is essential to prioritize the well-being and conservation of butterfly species over personal interests. Collection of butterflies from protected areas or private properties without permission from the competent authority is unethical. It is also crucial to avoid damaging butterfly habitats or disrupting their natural behavior. Collecting rare or endangered species should be avoided. In the Indian Biological Diversity Act 2002 are laid down the rules for conservation of biodiversity and sustainable use of its components. The State Biodiversity Authority of Assam acts by complying with Indian Biodiversity Authority for protection of Biodiversity of Assam. By adhering to ethical guidelines, we can ensure that the butterfly collection contributes positively to their conservation. In recent years, there has been an increasing emphasis on practicing non-invasive techniques that prioritize the well-being and conservation of butterflies and other wildlife. Here are some alternative methods for observing and studying butterflies without negative effect on natural balance:

Photography

Using a camera with a macro lens or a high-quality smartphone camera, one can capture detailed images of butterflies in their natural habitat. Photographing butterflies allows for detailed observation and documentation of their characteristics, behavior, and habitat preferences. The butterflies should be approached patiently, slowly and calmly to avoid scaring them away. Photographing butterflies can be used in citizen science projects and educational resource creations.

Binoculars and Spotting Scopes

Binoculars or spotting scopes may be used to observe butterflies from a distance without interference. This method is particularly useful for observing butterflies at a distance or in hard-to-reach areas.

Butterfly Houses and Gardens

Butterfly-friendly environments in the form of garden or designated butterfly house may be created by planting nectar-rich flowers and host plants that naturally attract butterflies. In a controlled setting, they can be observed without the need of capturing them. Butterfly houses can provide opportunities for long-term observation on biology and behaviour. They also provide a site for outdoor learning experiences for children and students.

Catching the butterfly: The butterflies should be observed to learn about their behavior, preferred habitats, and identifying features before attempting to catch them.

The butterfly net is used to catch the insect and one should be careful not to damage the wings or body during capture. The captured butterfly can be placed into the killing jar where the fumes will quickly and painlessly euthanize it. The butterfly is left in the jar for a few minutes to ensure that it is fully deceased.

Preserving and Mounting Butterflies

Once the butterflies are collected, mounting and preserving them properly is crucial to maintain their beauty and integrity. The mounted butterflies are stored in a clean and dry environment to prevent deterioration.

Pinning the butterfly: During pinning, the butterfly is gently spread out and pinned onto a mounting board. It is vital to handle the butterfly with care to prevent any damage to the wings or body. Entomological forceps are used to hold the specimen's thorax and gently insert a pin to the right of the mid-centre of the thorax.

Spreading the wings: The wings of the butterfly are spread carefully using the spreading board and pins. The wings are spread out to keep in a symmetrical posture exposing the wing pattern. The posture is maintained by inserting pins on the boundary of wings or using fine paper strips. This step requires patience and precision.

Drying the specimen: The pinned butterfly is then left to dry in butterfly boxes kept in environment controlled chambers with low humidity and temperature at least for a week.

Labeling: While the butterfly dries, essential data like the date, location, and habitat of collection, as well as any relevant notes about the specimen should be recorded.

Mounting: Once the butterfly is thoroughly dry, the specimen is gently placed on a mounting paper or foam.

Storage: The mounted butterfly is stored in an insect display case, away from direct sunlight and in a controlled environment to prevent microbial attack. The bodies of butterflies serve as good food for many insects. Naphthalene balls may be placed to avoid attacks of ants or other insects. Periodic monitoring is essential for prevention of damage.

Displaying and Showcasing Butterfly Collections

A well-curated butterfly collection can be a visually stunning display while also serving as an educational tool. When showcasing butterfly collections, factors such as lighting, temperature, and humidity should be considered. Display cases with glass fronts are ideal for protecting the specimens from dust and damage. The butterflies may be arranged in a visually appealing manner, considering their colors, sizes, and patterns. Each specimen must be labeled with its scientific name, common name, and relevant information.

Method for observation of characters

Descaling of butterfly wings : To study the venation pattern of wing descaling becomes necessary in case of some butterflies. The following procedures can be used.

Procedure 1: Using Ethanol

- Materials required: 70% isopropyl alcohol, a pair of fine-tipped forceps or tweezers, a soft-bristled paintbrush, and Q-tips.
- The butterfly is placed and fixed to a thermocol board by pinning .
- Using a soft-bristled paintbrush, the dry wings are first soaked in alcohol and gently brushed. The process is done by starting at the base of each wing and brushing toward the tip. Damaging the body parts should be avoided.
- The loosely attached scales come out easily with the strokes of the brush. The remaining scales are removed using a Q-tip in the same way.
- The wings are placed on a clean, dry surface and then allowed to air dry completely. During drying, touching the wings should be avoided as this could cause damage.

Procedure 2: Using KOH

- A 10% solution of potassium hydroxide (KOH) is prepared by dissolving 10 grams of KOH pellets in 100 milliliters of distilled water (≈60°C-80°C). Gloves are to be used while handling KOH, as it can cause burns and damage.
- The butterfly specimen is placed in a petri dish containing KOH solution so that the wings remain fully submerged in the solution. The specimen is kept in KOH for several hours or overnight depending on the size of specimen and extent of scaling present.
- Once the scales are loosened, the specimen is carefully removed from the KOH solution using forceps or tweezers. The specimen is placed in a dish containing distilled water and gently agitated to remove any remaining scales.
- The specimen is then rinsed several times to remove all traces of KOH and blotted dry using paper towel or soft cloth, taking care not to damage the delicate wings.

Note: KOH can make the wings brittle and fragile, so it is important to handle the specimen gently throughout the descaling process.

Study of genitalia: For the study of genitalia, the system has to be dissected out of the abdomen under a Binocular microscope. Before dissection of the genitalia, the abdomen is separated from the thorax and kept in 10% KOH overnight or heated without boiling for an hour to dissolve the fat and soft tissues of the abdomen. Once done, the abdomen is washed under tap water. For dissection, the abdomen is placed in dissecting plate by keeping the ventral side up. By soft pressing, the excess water and dissolved tissues can be removed. Angled pins, fine needles and forceps are used in dissection.

List of butterflies with conservation status

Comprehensive list of butterflies with IWPA conservation status and North east India status

Common Name	Scientific name	IWPA	NE Status	Photo contributor
Hesperidae				
Common Snow Flat	*Tagiades japetus*	Schedule II	-	Pooja Nag
Common Spotted Flat	*Celaenorrhinus leucocera*	Schedule II	-	Amrita Mech
Dark Branded Swift	*Caltoris brunnea caere*	NP	Endemic	Dipsikha Bora
Dark Velvet Bob	*Koruthaialos butleri*	NP	Endemic	Dipsikha Bora
Great Swift	*Pelopidas assamensis*	Schedule IV	-	Amrita Mech
Tree Filter	*Hyarotis adrastus*	Schedule IV	-	Amrita Mech
Veined Palmer	*Hidari bhawani*	NP	Endemic	Amrita Mech
Papilionidae				
Common Mime	*Papilio clytia clytia*	Schedule-II	-	Dipsikha Bora
Great Zebra	*Graphium xenocles*	NP	Endemic	Pooja Nag
Khasi golden birdwing	*Troides aeacus*	Schedule II	-	Pooja Nag
Khasi Yellow Helen	*Papilio nephelus chaon*	NP	Endemic	Dipsikha Bora
Oriental common bluebottle	*Graphium sarpedon*	Schedule II	-	Dipsikha Bora

Patkai Common Raven	*Papilio castor castor*	NP	Endemic	Phalgun Chetia
White Dragontail	*Lamproptera curius*	NP	Endemic	Phalgun Chetia
Nymphalidae				
Assam Unbroken Sergeant	*Athyma pravara*	Schedule II	Endemic	Dipsikha Bora
Autumn Leaf	*Doleschallia bisaltide*	Schedule II	-	Dipsikha Bora
Black Rajah	*Charaxes solon*	Schedule II	Endemic	Dipsikha Bora
Common Baron	*Euthalia aconthea sudhudana*	Schedule II	C	Amrita Mech
Common Nawab	*Polyura athamas*	Schedule II	-	Phalgun Chetia
Danaid Eggfly	*Hypolimnas misippus*	Schedule II	C	Dipsikha Bora
Gaudy baron	*Euthalia lubentia*	Schedule IV	-	S Islam
Great Archduke	*Lexius cyanipardus cyanipardus*	Schedule II	Endemic	Pooja Nag
Himalayan Grey Count	*Tanaecia lepidea lepidea*	Schedule II	C	Dipsikha Bora
Khasi Dark Archduke	*Lexius dirtea*	Schedule II	Endemic	Pooja Nag
Magpie Crow	*Euploea radamanthus*	Schedule IV	-	Pooja Nag
Myanmarese Wizard	*Rhinopalpa polynice*	Schedule II	Endemic	Dipsikha Bora Niranjan Barua
Peal's Palmfly	*Elymnias pealii*	Schedule I	Endemic	Phalgun Chetia
Plain Earl	*Tanaecia jahnu jahnu*	NP	Endemic	Dipsikha Bora

Plain Sailer	*Neptis cartica*	NP	Endemic	Dipsikha Bora
Spotted Palmfly	*Elymnias maleas*	Schedule II	-	Amrita Mech
Striped Blue Crow	*Euploea mulciber*	Schedule IV	-	Pooja Nag
Yellow Rajah	*Charaxes marmax marmax*	ScheduleII	-	Dipsikha Bora
Yellow Tipped Archduke	*Adolius pardalis*	Schedule II	-	Dipsikha Bora
Pieridae				
Chocolate Albatross	*Appias lyncida eleonara*	Schedule II	-	Niranjan Baruah
Common Gull	*Cepora nerissa*	Schedule II	NC	Dipsikha Bora
One-spot Grass Yellow	*Eurema andersoni*	Schedule II	-	Dulen Bora
Red Spot Jezebel	*Delias descombesi descombesi*	NP	Endemic	Phalgun Chetia
Sylhet Common Albatross	*Appias albina darada*	Schedule II	Endemic	Phalgun Chetia
Tree Yellow	*Gandaca harina assamica*	NP	Endemic	Dipsikha Bora
Lycaenidae				
Common Onyx	*Horaga onyx onyx*	Schedule II	-	Phalgun Chetia
Common Pierrot	*Castalius rosimon rosimon*	NP	C	Dipsikha Bora
Dark Pierrot	*Tarucus ananda*	Schedule IV	-	Phalgun Chetia
Himalayan Common Hedge Blue	*Acytolepis puspa gisca*	N EV	-	Phalgun Chetia

Naga Saphire	*Heliophorus kohimensis*	Schedule II	C	Dipsikha Bora
Oriental Apefly	*Spalgis epeus epeus*	NP	Not common	Pooja Nag
Oriental Orchid Tit	*Chliaria othona othona*	Schedule I	-	Phalgun Chetia
Pea Blue	*Lampides boeticus*	Schedule II	-	Pooja Nag

References

Govt. of India, Indian Wild Life Protection Act (1972).

Haribal, M. (1992). The butterflies of Sikkim Himalaya and their natural history. Sikkim Nature Conservation Foundation. Sikkim.

Kunte, K., Sondhi. S, and Roy, P. (Chief editors). Butterflies of India,v. 4.12. Indian foundation of butterflies. URL:https://www.ifoundbutterflies.org.browsing 2024

Neog, S. (2015). Butterflies of Gibbon Wildlife Sanctuary. Bhabani Books, Guwahati

Norman, T. (1953). Notes on the Lepidoptera of Assam-II. Some additions to Assam list and other notes. Journal of the Bombay Natural History Society, 51, 515-517

Saikia MK and Saikia, PK (1922). Butterflies Gauhati University-781014, Assam ISBN:178-81956086-6-0

Singh, A.P. (2015). Recent record of a rarely recorded species, the Veined Palmer Hidari bhawani de Nicéville, 1888 (Lepidoptera: Hesperiidae: Aeromachini) from Jorhat, Assam, India. Journal of Threatened Taxa, 7(1), 6839-6840, http://doi. org/10.11609/JoTT. o4031. 6839-40

Singh, A.P. (2020) Butterflies of Eastern Assam, NE India. Journal of Threatened Taxa,9(7), 10396-1042. DOI: 10.11609/jott.3177.9.7.10396-10420

Singh, A.P., Gogoi, L and Sebastain, J (2015). The seasonality of butterflies in a semi-evergreen forest: Gibbon Wildlife Sanctuary, Assam, northeastern India. Journal of Threatened Taxa, 7(1), 6774- 6787. http://doi.org/10.11609/JoTT. o3742.6774-87

Epilogue

The book's most important aspect is the incorporation of list of butterflies, along with identification criteria, specifically focusing on species accessible to the inhabitants of Assam. This collection also includes butterflies sighted in the fringe areas of several protected forests within the state.

Visiting well-maintained butterfly collections in museums and educational institutions has always left a lasting impression on me. Maintaining a butterfly collection requires infrastructure with controlled environmental conditions. Last year, in 2023, I had the opportunity to learn the history behind insect collection and preservation at the Smithsonian National Museum of Natural History in Washington, D.C. My visit to Malaysia earlier in 2012 took me to the Kuala Lumpur Butterfly Park, covering an area of about 80,000 square feet. The park is enriched with various components of a rich terrestrial ecosystem, including flowering plants, an aquatic body, and different species of ferns. It is maintained as a natural habitat for various invertebrates as well as lower vertebrates like fish and turtles, and houses about 5,000 butterflies showcasing the magnificence of their forms and patterns. It is a site for on-field education. The Universiti Kebangsaan in Malaysia is maintaining a remarkable collection of insect fauna, including butterflies of South East Asia. There are museums across the globe where insects are preserved time series wise. They are invaluable for studies on natural history. Several of my such visits within India and abroad, combined with my teaching experience in entomology at Dibrugarh University, Assam inspired my endeavour to give shape to the current book.

In 1963, meteorologist Edward Lorenz reported in his publication that small changes could have unimaginably large effects on the weather, a concept now known as the "Butterfly Effect." This theory suggests that a single flap of a butterfly's wings can eventually cause a tornado, challenging the classical understanding of nature. According to Lorenz, everything in nature is unpredictable and depends on its initial state.

The book, with its collection of butterfly photographs, will serve as a database of species commonly sighted in Assam over the last and the current decade. Since butterflies are excellent environmental indicators, this book will also serve as baseline of natural history for future studies on the effects of climate change in Assam in the years to come.

Abbreviations and Glossary

Adecticus	Pupa lacking functional mandible
Androconia	Scent scales on a male butterfly
Antennae	Pair of long jointed structure arising from head directed forward
Apex	Anterior tip of the wing
Apical area	Area adjacent to outer margin of wing
C	Common
Cell	Area of the wing enclosed by veins
Chrysalis	Butterfly pupa
Costa	Fore margin of the wing between the base and apex
Dcv	discocellular vein, the transverse veing forming the distal margin of cell in wing
Discal	Central area of the wing
Discal Band	The band located posterior to cell end bar
Dorsal	Upper surface
Dorsum	Inner margin of wing
DSF	Dry season form
Endemic	Present in a particular region
Epiculus	Tapering apical portion of antennal club
F	Female
IWPA	Indian wild life protection Act 1972
Lateral	On the sides

M	Male
Mya	Million years ago
N Ev	Not evaluated
NC	Not common
Obtect	Pupa in which appendages are glued to the body wall are called obtect pupa
Ocellus	Eye like pattern with pupil inside
Proboscis	Coiled tube bearing food canal
R	Rare
Scales	Modified flattened hairs covering the membranous wing
Tail	Slender hindwing projection covered with scales
Tarsus	The last division of leg bearing claws
Termen	Outer margin of wing
Tornus	Angle of wing between termen and dorsum
Tubercle	Small projection of skin
UNF	Underside of fore wing
UNH	Underside of hind wing
UPF	Upperside of forewing
UPH	Upperside of hind
V	vein
Ventral	Lower surface
Vs	vein space
WSF	Wet season form

List of Photo Plates of Butterflies

Hesperidae

Papilionidae

22	Khasi Yellow Helen, *Papilio nephelus chaon* Westwood	75
23	Lime Butterfly, *Papilio demoleus* Linnaeus	70
24	Oriental Common Bluebottle, *Graphium sarpedon sarpedon* Linnaeus	61
25	Oriental Tailed Green Jay, *Graphium agamemnon* Linnaeus	59
26	Patkai Common Raven, *Papilio castor castor* Westwood	73
27	Red Helen, *Papilio helenus* Linnaeus	76
28	White Dragontail, *Lamproptera curius Fabricius*	58

Nymphalidae

1	Angled Castor, *Ariadne pallidior* Frusthofer	133
2	Assam Unbroken Sergeant, *Athyma pravara* Fruhstorfer	110
3	Black Rajah, *Charaxes solon solon* Fabricius	51
4	Blue Glassy Tiger, *Ideopsis similis* Linnaeus	86
5	Blue King Crow, *Euploea klugii klugii* Moore	87
6	Chinese Lemon Pansy, *Junonia lemonias lemonias* Linnaeus	123
7	Chocolate Pansy, *Precis iphita* Cramer	125
8	Colour Sergeant, *Athyma nefte inara* Westwood	110
9	Commander, Limenitis *procris procris* Cramer	109
10	Common Baron, *Euthalia aconthea sudhudana* Cramer	116
11	Common Castor, *Ariadne merione assama* Evans	132
12	Common Evening Brown, *Melanitis leda* Linnaeus (DSF)	90
13	Common Evening Brown, *Melanitis leda* Linnaeus (WSF)	90
14	Common Five Ring, *Ypthima baldus baldus* Fabricius	96
15	Common Five Ring, *Ypthima baldus* Fabricius (DSF)	98
16	Common Four Ring, *Ypthima hubenri* Kirby	95
17	Common Jester, *Symbrenthia hippoclus* Cramer	129
18	Common Lascar, *Pantoporia hordonia hordonia* Stoll	115

Pieridae

Lycaenidae

Index of Common Names of Butterflies

Hesperidae

Papilionidae

Nymphalidae

Pieridae

Lycaenidae

Index of Scientific Names of Butterflies

Nymphalidae

8	*Athyma pravara* Fruhstorfer, 1906	110
9	*Athyma selenophora* Kollar, 1848	112
10	*Cethosia cyane* Drury, 1770	102
11	*Charaxes solon solon* Fabricius, 1793	120
12	*Charaxes marmax marmax* Westwood, 1848	119
13	*Cirrochora aoris* Doubleday, [1847]	105
14	*Cyrestis thyodamas* Boisduval, 1836	131
15	*Danaus chrysippus chrysippus* Linnaeus, 1758	82
16	*Danaus genutia genutia* Cramer, [1779]	82
17	*Doleschallia bisaltide indica* Moore[Cramer, 1777]	128
18	*Elymnias hypermnestra undularis* Drury, 1773	99
19	*Elymnias malelas* Hewitson, 1863	100
20	*Elymnias pealii* Wood-Mason, 1883	100
21	*Euploea core* Cramer, [1780]	88
22	*Euploea klugii klugii* Moore, [1858]	87
23	*Euploea mulciber* Cramer, [1777]	89
24	*Euploea radamanthus* Fabricius, 1793	88
25	*Euthalia aconthea sudhudana* Cramer, 1779	116
26	*Euthalia aconthea garuda* Moore, 1857	117
27	*Euthalia lubentia* [Cramer, (1777)]	208, 236
28	*Hypolimnas misippus* Linnaeus,1764 (M)	127
29	*Hypolimnas bolina* Linnaeus, 1758 (F)	126
30	*Hypolymnas bolina* Linnaeus, 1758 (M)	127
31	*Ideopsis similis* Linnaeus, 1758	86
32	*Junonia almana* Linnaeus, 1758	124
33	*Junonia atlites* Linnaeus, 1763	123
34	*Junonia lemonias lemonias* Linnaeus, 1758	123
35	*Lexius cyanipardus* Butler, 1868	108
36	*Lexius dirtea khasiana* Swinhoe,[1890] (= *intermedia* Tyt)	106
37	*Lexius pardalis* Moore, 1878	107
38	*Limenitis procris procris* Cramer, 1779	109
39	*Melanitis leda leda* Linneaus, 1758	91
40	*Melanitis leda* Linnaeus, 1758 (DSF)	90

Blue Tiger, *Tirumala limniace*

About the Authors

The author, **Dr. (Mrs.) Dipsikha Bora**, has been teaching **master's courses** in Life Sciences at Dibrugarh University, Assam, India, for over three decades in the areas of Entomology, Endocrinology, Toxicology, and Bioethics. Her research interest is focused on bioresources, insect-plant interaction and plant based pesticides. The author has contributed to her field of research with more than 60 publications in peer-reviewed journals and 10 book chapters. She has supervised 15 Ph.D. theses under Dibrugarh University and is granted with two Indian patents . The author has successfully completed five extramural research projects funded by UGC, DST, DBT, Govt. of India, as the Principal investigator. She delivered talks in several national and international seminars/conferences and visited Amsterdam, Malaysia, Thailand, Florida, Maryland and Washington DC, for academic purposes. She is also a member of several professional bodies.

Anisha Phukan, DST, WISE Ph. D Fellow(KIRAN) obtained her **Master's** degree in **Life Sciences** (Zoology) from Dibrugarh University and is currently pursuing her research in the Department of Life Sciences, Dibrugarh University. She is specialized in **Entomology.**

Dr. Amrita Mech is currently serving as an Assistant Professor in the Department of Zoology, Sonari College, Charaideo, Assam. She obtained her Ph.D. under the Department of Life Sciences, Dibrugarh University. Her areas of research interest include the study of insect diversity, biopesticides, environmental biology, and insect-host plant interaction. She has eight research papers, book chapters and two edited books to her credit.

Phalgun Chetia completed his **Ph.D.** under Dibrugarh University in 2023. His interest in wildlife started by freelancing in conservation drives in Kaziranga Tiger Reserve since 2008. He was a researcher in the "Indo-African Animal Exchange Program" under **MoEF,** Government of India, in 2009. Later, he focused on arachnids around Kaziranga Tiger Reserve. He has contributed **to the documentation of** spiders of Upper Assam on behalf of the Assam Biodiversity Board.

Pooja Nag has completed her **post-graduation** from Dibrugarh University in **Life Sciences** (Zoology) with **a specialization in Entomology**.

Jyoti Agrahari completed Masters from Royal Global University,Guwahati, Assam and currently pursuing research in the Department of Life Sciences, Dibrugarh University

Gaudy baron, *Euthalia lubentia* basking
Protected under Schedule IV of IWPA, 1972

Nectaring Grass Demon, *Udaspes folus*

Pupae formation in line by Grass yellow

Fruit bait

Mud puddling by the Swallowtails after the rains

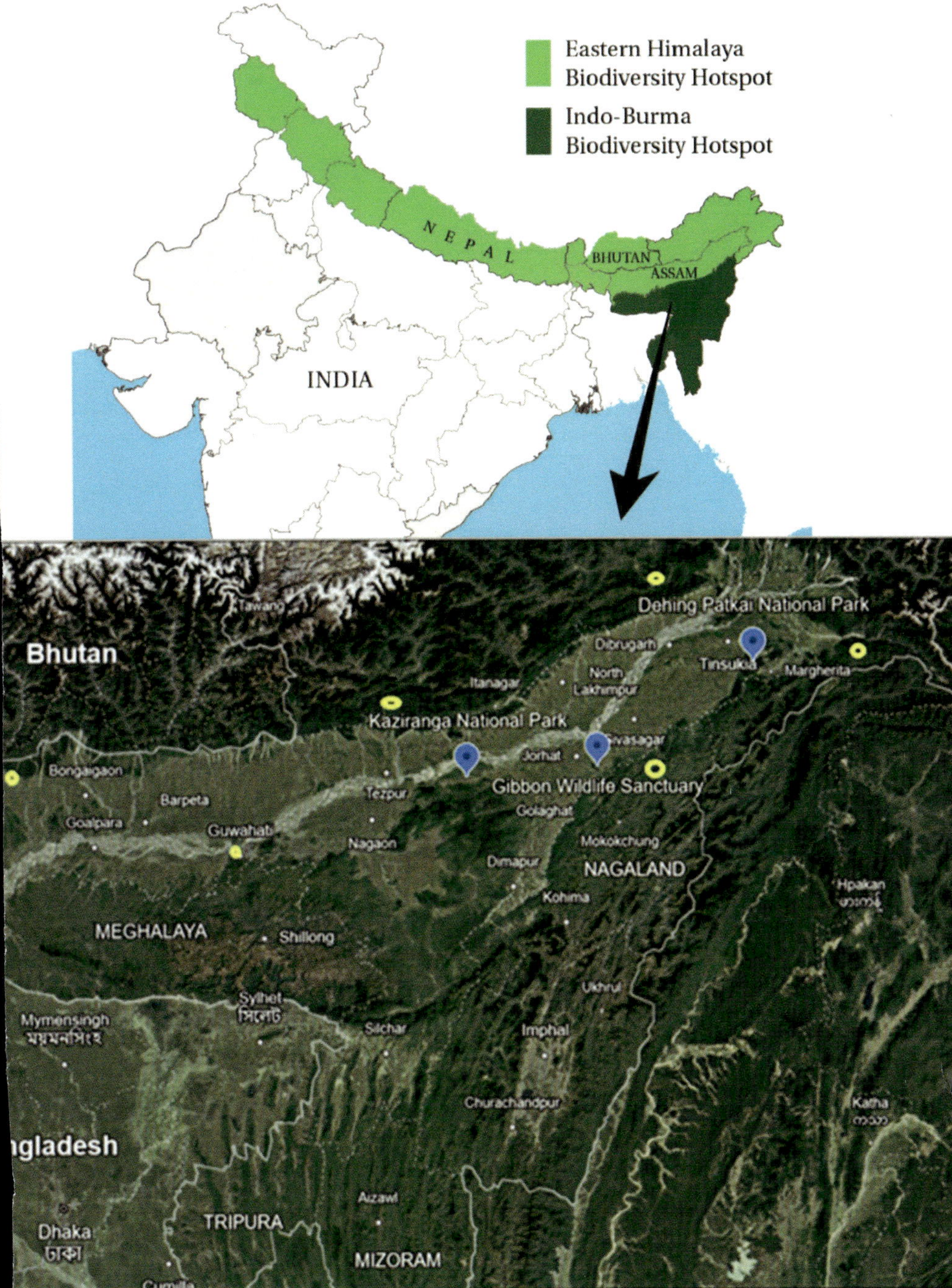

Map of study site is adopted from Google Earth Pro.

Notes